MAPPING INVISIBLE WORLDS

COSMOS
The Yearbook of the Traditional Cosmology Society

Volume 9
1993

MAPPING INVISIBLE WORLDS

edited by
GAVIN D. FLOOD

General Editor of Cosmos: Emily Lyle

EDINBURGH UNIVERSITY PRESS

Advisory Board

Professor Paul Badham, Theology and Religious Studies, University of Wales, Lampeter
Lance Cousins, Comparative Religion, Manchester University
Professor Anna L. Dallapiccola, Fine Art, University of Edinburgh
Dr Charles Jedrej, Social Anthropology, University of Edinburgh
Dr Glyn Richards, Religious Studies, Stirling University
Dr Jacqueline Simpson, The Folklore Society
Professor Emeritus Cyril Williams, Theology and Religious Studies, University of Wales
Rosemary Muir Wright, History of Art, University of St Andrews

IN MEMORY OF DEIRDRE GREEN

© Edinburgh University Press, 1993

Edinburgh University Press Ltd
22 George Square, Edinburgh

Set in Linotron Times Roman
by Photoprint, Torquay, and
printed in Great Britain
by Hartnolls Ltd, Bodmin

A CIP record for this book is available from the British Library

ISBN 0 7486 0470 7

Contents

GAVIN D. FLOOD
Introduction   1

C. R. A. MORRAY-JONES
The 'Descent to the Chariot' in Early Jewish Mysticism
and Apocalyptic   7

WENDY PULLAN
Mapping Time and Salvation: Early Christian Pilgrimage
to Jerusalem   23

J. McKIM MALVILLE   JOHN M. FRITZ
Mapping the Sacred Geometry of Vijayanagara   41

EMILY LYLE
Internal–External Memory   63

MARK NUTTALL
Place, Identity and Landscape in North-west Greenland   75

GRAHAM HARVEY
Gods and Hedgehogs in the Greenwood: Contemporary
Pagan Cosmologies   89

HILDA ELLIS DAVIDSON
Mythical Geography in the Edda Poems   95

URSZULA SZULAKOWSKA
The Pseudo-Lullian Origins of George Ripley's Maps and
Routes as developed by Michael Maier   107

CYRIL WILLIAMS
Waldo Williams: A Celtic Mystic?   127

ROY WOODS
Against Mapping Invisible Worlds in Rilke's *Duino Elegies*   139

DAVID MACLAGAN
Inner and Outer Space: Mapping the Psyche 151

BRIAN BOCKING
'If you meet the Buddha on the map . . .': The Notion of Mapping Spiritual Paths 159

Notes on Contributors 163

GAVIN D. FLOOD

Introduction

The historical, social, and psychological sciences of the last two centuries have indeed demonstrated beyond the shadow of a doubt that every religious tradition is full of projections of a variety of very this-worldly human interests. . . . The modern scientific study of religion would not be possible without these demonstrations of mundane determinants of what purports to be extramundane. Let it be stipulated in principle . . . that these demonstrations are valid. The point is, quite simply, that this is not the whole story. An analogy may be helpful here. Take the case of travellers returning home with accounts of a faraway country. Assume that it can be demonstrated beyond the shadow of a doubt that every one of these accounts is determined by the historical, socioeconomic, and psychological characteristics of the traveller in question. . . . As the critical observer analyses all these reports, it is perfectly plausible for him to perceive the faraway country as a gigantic projection of the observer's own country. Indeed, the travellers' accounts will be very useful in gaining a better understanding of their home country. None of this, however, invalidates the proposition that the faraway country does indeed exist and that something about it can be gleaned from the travellers' accounts. The final point is not that Marco Polo was an Italian – and, who knows, an Italian with all sorts of class resentments and with an unresolved Oedipus complex – *but that he visited China*. (Berger 1979: 122–3)

This quotation was the jumping-off point for a conference of the Traditional Cosmology Society held at St David's University College, Lampeter, in September 1992.[1] The quote served to focus the interests of scholars from a wide variety of disciplines, including English Literature, Folklore Studies, Religious Studies, German Studies, Fine Art and even Astrophysics. What the participants had in common was an interest in the ways in which human communities and individuals, from a variety of times and places, have represented their values, aspirations and beliefs in the nature of themselves and their worlds. More specifically, the participants shared an interest in representations which claim to be, or can be viewed as, 'maps' of those diverse worlds. These representations or maps – such as a tradition's sacred geography, its mythology, its art and its spiritual paths[2] – may express a tradition's transpersonal realities: its source, gods and non-material worlds. A 'map' might also represent an individual's private, 'inner world', or, indeed, make a statement that such 'worlds' cannot be mapped.

Many of the representations of 'invisible worlds' discussed in this volume serve to illustrate the fact that meaning in pre-modern cultures is a function of a person's location within a cosmos. Maps of invisible worlds serve to show a community or individual where they are in relation to a wider context or totality and, by implication, where they should or should not be going. Such maps provide moral resources through which communities regulate their affairs and from which they draw their inspiration. This might be contrasted with contemporary 'western' or 'de-traditionalised' communities and individuals, where maps of invisible worlds are maps of the 'internalised self' (Taylor 1989: 456–66) and of the private citizen.

The chapters in this book, which are a representative selection of papers given at the conference, deal with a variety of themes and cover wide conceptual, temporal and geographical areas. They fall roughly into four categories: those examining the ways in which traditions map their spiritual paths, and the problems of mapping such paths (Morray-Jones and Bocking); those examining some of the ways in which mythical worlds are mapped with specific geographies (Davidson and Szulakowska); those dealing with how the physical world can be imbued with mythological or religious meaning (Pullan, Malville and Fritz, Lyle, Nuttall and Harvey); and lastly, examples of mapping private worlds in modern, western visual art and poetry, which may or may not have objective correlates (Williams, Woods and Maclagan). There is, of course, much overlap between these categories, and various themes echo in different chapters. For example, Bocking underlines the problems of reifying maps of invisible worlds, a point echoed by Maclagan and Woods.

The mapping of mythical worlds or cosmologies often corresponds to the mapping of spiritual paths; a journey 'within' can also be regarded as a journey through the cosmos. This idea, or even principle of esoteric traditions, is illustrated in the first chapter, by Chris Morray-Jones, which examines the visionary literature of Jewish *Merkabah* mysticism. Morray-Jones begins by defining his use of 'mysticism' as the cultivation of altered states of consciousness or ecstasy, and the 'mystical tradition' as the context in which these experiences occur. This context is the vision of God upon the Throne of Glory, beheld by the mystic in experience, which is also a journey into the body and into the cosmos. While the *Merkabah* mystic maps the body of God within his own body, the early Christian tradition mapped the heavenly city of Jerusalem upon the earthly city. Wendy Pullan's chapter, which follows, shows how Jerusalem was revered as a holy city by Christians from as early as the fourth century CE and how the Christian's pilgrimage to Jerusalem was a symbol of the soul's journey to salvation. This idea of a city as a projection or, rather, reflection of a sacred reality is further illustrated by McKim Malville and John Fritz. They show how the structure of the medieval Hindu city of Vijayanagara is based on Hindu cosmological principles and reflects the social or power structures of Hindu culture, centred upon the ideal of the king.

Journeying south and east we move to the Caroline Islands which are the focus for much of the discussion of Emily Lyle's chapter. She demonstrates how the visible world can function as an external memory, or 'symbolic storage system'. Taking illustrations from Micronesian navigation, she shows how complex journeys could be made with the use of 'maps' incorporating imaginary places. The overlaying of geographical locations with symbolic meaning is further illustrated by Mark Nuttall, who describes how place is infused with meaning for the seal-hunting communities of north-west Greenland. Indeed, the naming of places, particularly with personal associations, is a way of articulating cultural identity. The idea of the 'natural world' (which itself is, surely, a cultural construction and is not pre-given) being imbued with cultural meanings can also be seen, as Graham Harvey shows, in contemporary Paganism. Here the ideal of the 'Greenwood' is both a place and an invisible world populated with visible and invisible entities.

Staying with invisible 'pagan' worlds, Hilda Davidson describes the Viking mythology expressed in the Edda poetry of Norway and Iceland. In examining these different worlds inhabited by a variety of beings, she illustrates how the Norse mythological cosmologies are not static, but fluid and vital. Urszula Szulakowska then examines very different kinds of invisible worlds in her chapter on alchemy, showing how the alchemical process described by the fifteenth-century alchemist, George Ripley, was elaborated by the seventeenth-century work of Michael Maier.

It could be said that Maier's illustrations of the alchemical process were also illustrations of his own invisible worlds. The next three chapters deal more specifically with the public expression of private worlds in western poetry and visual art in the twentieth century. Cyril Williams places the Welsh poetry of Waldo Williams in the context of modern discussions about mysticism, showing how the poet attempted to express a powerful and distinctive experience which may yet have parallels in other texts and traditions. This theme of private experience and its public expression is taken up by Roy Woods in his exposition of Rilke's poetry and Rilke's translation into words – or mapping – of an immediately apprehended 'non-dual space'. With reference to the visual arts, David Maclagan then examines how subjective maps, referring to personal experiences, emotions and dreams, can form the basis of a personal mythology which, he argues, is a more appropriate expression of inner worlds than psychological maps claiming objectivity. Our concluding chapter by Brian Bocking reflects these concerns, and questions the usefulness and, indeed, the very basis of spiritual maps. He reflects on the nature of maps and suggests that, rather, it is often the guide who is regarded as more significant.

In the end we are still left with the problem implied by Berger's quote, and partly addressed in some of the chapters concerning the existence of 'China'. His point echoes that of Smith, quoting Alfred Korzybski, that 'map is not territory' (Smith 1978: 309). However, unlike Smith, who thinks that 'maps are all we possess', Berger implies that we could, in principle, visit 'China'. One of the problems here is the complexity and inevitable incongruity of the maps involved.

With the development of modernity and the decline of any external or objective moral source, we can see the erosion of a shared idea of objective, invisible worlds and the development of an internalised view of the self in which invisible worlds are specific and private. Inwardness, and the 'mapping' of this inwardness through the particular creative imagination, has become, as Taylor has shown, part of the modernist sensibility (Taylor 1989: 481). This may be contrasted with more traditional religious views, and the views of pre-modern societies, in which the invisible worlds of narrative traditions and esoteric teachings extend beyond the inwardness of any particular individual. Most religious traditions (or traditions' Marco Polos) would endorse a view of the reality of the invisible worlds far beyond those of 'average' human experience, and existing independently of human projections. Some esoteric traditions, on the other hand (I'm thinking here particularly of certain forms of Buddhism), might veer away from both the objective inner cosmos and the private invisible worlds of the internalised self. Such a tradition might say that to experience 'China' is, in fact, simply to be where we are; a sentiment perhaps echoed in the famous lines of T. S. Eliot:

> We shall not cease from exploration
> And the end of all our exploring
> Will be to arrive at where we started
> And know the place for the first time.
> (Eliot 1972: 59)

The problem will, of course, remain.

NOTES

1 The quote was supplied and circulated by Dr Emily Lyle. The foundations for the conference were laid at an earlier conference on Mysticism held in Oxford in 1990. The planning for this was undertaken by Dr Lyle in conjunction with the late Dr Deirdre Green, to whose memory the conference and the volume are dedicated.
2 I take the terms 'spiritual path' in a purely descriptive sense to mean a soteriological system of doctrines, techniques or practices, usually involving the presence of a teacher or conveyor of a tradition. My use is akin to Sherry's term 'spiritual life', by which he means 'a way in which people attempt to acquire holiness and an awareness of the presence of God through prayer, meditation and other devotional practices' (1984: 3).

REFERENCES

Berger, P. (1979). *The Heretical Imperative*. New York: Anchor Press/Doubleday.

Eliot, T. S. (1972). *The Four Quartets*. London: Faber.

Sherry, P. (1984). *Spirits, Saints and Immortality*. London: Macmillan.

Smith, J. Z. (1978). *Map is Not Territory. Studies in the History of Religions*. Leiden: Brill.

Taylor, C. (1989). *Sources of the Self*. Cambridge: Cambridge University Press.

C. R. A. MORRAY-JONES

The 'Descent to the Chariot' in Early Jewish Mysticism and Apocalyptic

In recent years, *Merkabah* mysticism has attracted a good deal of attention within Jewish Studies, and a number of scholars have perceived its potential significance for our understanding of the origins of Gnosticism and Christianity. However, the material is not easily accessible to the non-specialist reader. Though the situation with regard to the textual sources is steadily improving, they are well-known only to a small number of specialists in early Judaica who have not yet reached a consensus with regard to their tradition–historical background. I propose here to offer a brief overview of the subject, and to indicate something of its importance for the study of Christian origins. First, however, I should define two terms.

To begin with, I do not wish to enter into a philosophical or theological debate as to what constitutes 'mysticism'. I propose to define this term (for the purpose of the following discussion) in purely pragmatic and minimalist terms: mysticism is the deliberate cultivation of states of 'altered consciousness', such as trance or ecstasy. Spontaneous experiences of such a kind do not fall within this definition: a mystical tradition is one in which practices for the obtaining of such experience (e.g. ascetic disciplines, theurgic rituals and/or auto-hypnotic techniques) are developed and handed down within a religious context. This context provides the mystic with an interpretative framework by which he can make sense of his experience, and which to some extent pre-determines that experience. In Jewish '*Merkabah* mysticism' this framework is provided by Ezekiel's description of his vision of God upon the heavenly Throne of Glory, which came to be called *hammerkabah* or 'the Chariot', augmented by other scriptural 'enthronement theophanies' such as Isaiah 6 and Daniel 7. The earliest instances of the term *Merkabah* in a relevant context are Sirach 49: 8 (of

Ezekiel's vision) and the 'Songs of the Sabbath Sacrifice' in the Dead Sea Scrolls (Newsom 1985), which are strikingly similar to passages in the later mystical texts. Note in passing that according to the first-century Jewish historian Josephus (*Jewish War* II.142), the Essenes (almost certainly the authors of the Scrolls) were believed to know the angels' secret names.

The adjective 'esoteric' refers to a body of secret, traditional teaching which is reserved to a circle of initiates. Not all mysticism is esoteric and not all esotericism is, properly speaking, mystical, but rather speculative and theosophical. However, esoteric teachings frequently have or claim for themselves a basis in mystical experience, and mystical techniques are often part of esoteric lore. Jewish mysticism has always been practised in the context of an esoteric tradition which was claimed to be a revealed wisdom handed down since the time of Moses, though augmented by subsequent revelations to mystical practitioners. The opening passage of the mystical compilation first published by Scholem (1965: 103–17), under the title *Ma'aseh Merkabah*, embodies this perspective very clearly (*Synopse* §544):

> R. Ishmael said:
> I asked R. Aqiba for the prayer that one should pray when he ascends to the *Merkabah*, and I requested from him the praise of *ROZAYA*', LORD God of Israel, since he knows what it is.
>
> He said to me:
> With sanctification and purity in one's heart, one should pray this prayer:
>
> Be Thou blessed forever upon the Throne of Glory,
> Thou Who dwellest in the chambers of heaven
> And in a glorious palace,
> For Thou hast revealed the secrets,
> And the secrets within the secrets,
> The concealed things
> And the things concealed within the concealed things,
> To Moses,
> And Moses has taught them to Israel
> That they might perform the Torah through them
> And increase, through them, the Talmud.

It can be seen that the writer, who evidently identified wholeheartedly with the values, community and traditions of Rabbinism, was concerned with the revelation of heavenly secrets on which depend the true interpretation and application of Torah. As in many Apocalyptic writings, this is closely associated with the

The 'Descent to the Chariot'

vision of God which the practitioner sets out to achieve. Indeed, they include the liturgical means whereby this vision is obtained.

Our first problem is how to place this esoteric–mystical tradition within our picture of Ancient Judaism as a whole. There are basically three bodies of evidence to be considered. The first of these is the Apocalyptic (and related) literature of the Second Temple and early Christian periods. Though the term *Merkabah* is nowhere employed in this literature, many Apocalypses describe heavenly journeys during which celestial secrets are revealed and which often, though not always, culminate in the vision of God's appearance in bodily form upon His Throne. All involve visions of angels. What we do not find in the Apocalypses is anything in the way of mystical instruction, and this has led some scholars (e.g. Halperin 1988) to believe that they are mere literary fantasies rather than products of a genuine visionary–mystical tradition. It is true that an Apocalypse is a narrative composition with a polemical agenda (which varies according to the sectarian bias of the writer) and that the descriptions of visions constitute, in effect, a claim that the book's message is authoritative revelation. However, this simply means that these works were intended for public circulation, to disseminate theological ideas, and that they were not meant to be books of mystical instruction. It does not exclude the possibility that they are derived from a genuine visionary–mystical tradition and several scholars have found indications that this is indeed the case (e.g. Rowland 1982).

Secondly, it is necessary to consider the rabbinic traditions about *Ma'aseh Merkabah* ('the work/story of the Chariot'), which are found in both talmudic and midrashic literature. In the *midrashim*, they are frequently associated with the Sinai theophany, and so with the revelation of the Torah. In this context belong the stories of Moses' ascent into heaven to receive the Torah, often in the face of angelic opposition. The talmudic sources contain two types of material. There is a genre of 'horror-stories' warning against involvement in *Ma'aseh Merkabah*, in which ill-advised individuals come to various kinds of sticky end (sudden death, madness, leprosy and so on). On the other hand, we find stories of great Rabbis who successfully 'expounded the *Merkabah*' and produced supernatural phenomena by so doing. Thus, these sources display an ambivalent attitude towards *Ma'aseh Merkabah*, and the overall impression is of something mysterious and wonderful, but terrifyingly dangerous and forbidden.

The third body of evidence to be considered is the collection of texts known as the *Hekhalot* (i.e. 'palaces' or 'temples') literature, which have survived mainly in manuscripts from circles of medieval German and eastern European mystics. The literary organisation of these strange, rambling compilations is extremely fluid, varying greatly between the manuscripts, and what is true of Rabbinic literature in general is much more so of these documents: the text, such as it is, represents a relatively late stage in the development of the tradition and evolved over a considerable period of time. The compilers claim that they are recording esoteric traditions handed down from the Rabbis of the first and second centuries CE. The accuracy or otherwise of this claim is still disputed. Scholem (1961, 1965), Gruenwald (1980, 1988) and others have argued that *Merkabah* mysticism was, indeed, a direct continuation of Apocalypticism within Rabbinic Judaism. Others have argued that the Rabbinic *Ma'aseh Merkabah* was a purely speculative tradition and that the ecstatic mysticism of the *Hekhalot* literature developed only in circles marginal to Rabbinism, in late and post-talmudic times (Halperin 1980, 1988; Schäfer 1984, 1986). My own analysis, however, supports a modified version of the Scholem–Gruenwald hypothesis (Morray-Jones 1988). The data suggest that esoteric and mystical traditions associated with the vision of God and His Throne were inherited from Apocalyptic circles, and enthusiastically developed, by some early Rabbis. Other Rabbis, however, were hostile towards these traditions, mainly because they were also being developed by groups whom they regarded as heretical (i.e. Christians and Gnostics). While it cannot be assumed that everything in the *Hekhalot* literature goes back to the first two centuries, the writers' claim to be the heirs to a tradition from this time and milieu deserves to be taken seriously.

Unlike the Apocalypses, the *Hekhalot* texts are concerned with mystical instruction. They are so called because they describe, and give instructions regarding, a visionary journey through seven concentric heavenly palaces (*hekhalot*) to behold the vision of God seated upon His Throne of Glory or *Merkabah*. The most complete account of this journey is given in *Hekhalot Rabbati*, where R. Nehunya b. HaQanah reveals the mystical method to his disciple R. Ishmael and 'the entire great and small Sanhedrin' who are assembled in the Temple. Nehunyah begins by describing a magical, and apparently auto-hypnotic, method of inducing trance (*Synopse* §§204–5):

> When a man wants to descend to the *Merkabah*, he should invoke *SURIYAH*, the Prince of the Countenance, and adjure him a hundred and twelve times by *TUTRUSIYA*, the LORD, who is called *TUTRUSIYA TZORTEQ TOTAR-*

The 'Descent to the Chariot'

BI'EL TOFGAR 'ASHERUYALAIYA ZEBODI'EL ZOHARARI'EL TĚNADA'EL SHEQEDHUZIYA 'ADHI-BIRON and 'ADIRIRON, the LORD God of Israel.

Let him not add to the hundred and twelve times, neither let him subtract therefrom! If he adds or subtracts, 'his blood is on his own head' (Joshua 2: 19)! Rather, while his mouth is pronouncing the names, let the fingers of his hands count one hundred and twelve times. Then he will descend and master the *Merkabah*.

Following this, he travels in a trance through the seven palaces and reveals, by automatic speech, the names of the terrifying angelic guardians of the gateways, who will only allow the traveller to pass if they are shown the correct magic seals, on which are inscribed the magical names of God. Finally, the visionary is admitted to worship before the *Merkabah* in the innermost palace. There follows a lengthy compilation of hymns which the mystic is instructed to recite, and with which the text (in its present form) ends.

The object of the climactic vision in the seventh palace, at the culmination of the heavenly journey, is the appearance of God as a vast and overpoweringly glorious human form of fire or light (Ezekiel's 'likeness of the appearance of a man') enthroned upon the *Merkabah*. A distinction (though, in the *Hekhalot* literature, no discontinuity of identity) is observed between this visible appearance and God as He exists in Himself beyond the seventh heaven, to which He descends to manifest in a visible form and receive the worship of His Creation. This form is referred to as 'the Glory' (*hakkabod*), frequently 'the Great Glory' (*hakkabod haggadol*), or 'the Power' (*haggeburah*). This terminology goes back to Ezekiel (1: 28 etc.), and is frequently found in Apocalyptic literature, where *kabod*, especially, is used as a technical term for the appearance of God on the heavenly 'Throne of Glory' (*kisse' hakkabod*). Both *kabod* and its Aramaic equivalent, *yeqara'*, are derived from verbal roots meaning 'to be heavy' and the Glory is, therefore, a materialisation of the Divine Essence in human form and/or as light. Consider in this light *Hekhalot Rabbati* 3.4 (*Synopse* §99):

And three times every single day, the throne of Thy Glory (*kisse' kebodka*) prostrates itself before Thee and says to Thee: 'O ZOHARARI'EL, LORD God of Israel, make Thyself heavy/glorious (*hitkabbed*) and sit upon me, O wondrous King, for the burden of Thee is lovely [some mss add:

'and precious (*yaqar*)'] upon me, and not heavy (*kabed*) upon me, as it is written: Holy! Holy! Holy is the LORD of Hosts!' It is clear that this passage preserves a very ancient idea indeed. Compare Paul's expression at 2 Corinthians 4: 17: '. . . an eternal weight of glory beyond all comparison.'

Attention should be drawn to the importance of praise in the *Hekhalot* literature. Long sections of these texts consist of grandiloquent hymns and prayers, which are sometimes said to have been learned from helpful angels. The final passage of *Ma'aseh Merkabah* (*Synopse* §§595–6) is a particularly fine example:

Rabbi Ishmael said:

I said to Rabbi Aqiba: How is one able to look upwards beyond the Seraphim that stand above the head of *ROZAYA'*, the LORD God of Israel?

He said to me:

When I had ascended to the first palace, I prayed a prayer, and I saw from the palace of the first firmament as far as the seventh palace. And as soon as I had ascended to the seventh palace, I pronounced the names of two angels, and I gazed upwards beyond the Seraphim. And these are they: *SYRD* and *HGLYN*. And as soon as I had pronounced their names, they came and took hold of me and said to me: Son of Man, do not be afraid. He is the Holy King Who is sanctified upon the High and Exalted Throne, and He is Excellent for ever and Majestic upon the *Merkabah*!

In that hour, I saw upwards beyond the Seraphim that stand above the head of *ROZAYA'*, the LORD God of Israel!

And what is this prayer?

1. Blessed art Thou, O LORD, the One God,
   The Creator of His World by His Name, which is One!
   The Maker of all things by a single command!
   In the heights of the heavens,
      Thou hast established Thy Throne,
   And Thy seat Thou hast placed on Thy heavenly peaks!
   Thou hast set Thy Chariot in the uppermost heaven!
   Thou hast planted Thine high country with Wheels of Glory!

2. Troops of fire praise Thy renown!
   Fiery Seraphim extol Thy praise!
   All of them pleading in a plaintive whisper,
   Uttering praise as they go to and fro,
   Walking in terror, fainting in fear,

The 'Descent to the Chariot'

Laden with glory to adorn the Maker of All!
They are full of eyes upon their rims,
Their appearance is as the appearance
   of the lightning-flash!
Their brightness is comely, their flavour is sweet,
And they rise up, side by side with each other,
Offering praise!

3. The pure Living Creatures rise up:
'HOLY! HOLY! HOLY!'
Say the ministering angels before Thee!
Their faces are the colour of the wheel of fiery heat!
Their brightness shines like the brilliance of the sky!
Their wings are extended, their hands are stretched
   forth,
The sound of their wings is like the sound of great
   waters!
Fiery torches flare continuously and issue forth
      from the wheels of their eyes!
With a voice like a great earthquake,
      they sing hymns before Thee!
They are filled with brilliance, they give forth brightness!
Adorned in a comely fashion, they go forth exulting
      and come in rejoicing,
In comely brightness before the Throne of Thy Glory!

4. In terror and in fear, they perform Thy Will,
Praising Thy Great, All Powerful and Revered Name,
Uttering praise and glorification
      for the remembrance of Thy Kingdom,
Shouting and jubilation – for there is none like Thee!
And there are none like Thy priests,
      and none like Thy saints!
And there is nothing like Thy Great Name for ever
      and through eternities of eternities!

5. Raging of the sea and drought are foreseen in the earth,
      and earthquake!
The whole universe will be shaken by Thy Power!
Thou wilt enliven the dead, and the dead shall arise
      from their dust!
Great is Thy Name for ever!
Noble is Thy Name for ever!
Holy is Thy Name for ever!
YHWH is One! *YHWH* is One!

> *YAH, YEHU* is Thy Name! *YHWH* is Thy Name for ever!
> *YHWH* is Thy Memorial from generation unto generation!
>
> 6. *SRTP Z'N Z'PY YH MQM' NQM NNQWN Y'RDD 'BG BG HWY HGG HG HW YW!*
> Everlasting is Thy Power! Everlasting is Thine Holiness!
> Everlasting is Thy Kingdom in the heavens and on the earth!
> Therefore let us call upon Thy Name! Let us bless Thy Power!
> Let us glorify and offer praise
> before the Throne of Thy Glory!
> For there is none like unto Thee
> in the heavens or upon the earth!
> Blessed art Thou, O LORD, the Eternal Rock!

A central theme of this text as a whole is the question: how can the mystic behold the vision of the Glory without being overpowered and destroyed, or consumed by fire? Hymns such as that shown above are given in response to this question, and we are told that if the mystic recites them in a state of ritual and moral purity they will provide him with theurgic protection and enlist the help of friendly angels. In this final passage of the text, we are taken one stage further: the mystic is enabled to look 'upwards beyond the Seraphim that stand above the head of *ROZAYA'*, the LORD God of Israel' – that is, into the secret dimensions of the Godhead beyond His manifestation as the Glory on the Throne. The image is derived from Isaiah 6: 2 (see below). Note the theurgic importance of the angels' names (in the second paragraph), knowledge of which enables the mystic to invoke their aid. The following prayer, which is actually the means of ascent, is a fine example of *Merkabah* imagery, and is packed with allusions to Ezekiel 1. Note that the vision is both glorious and joyful, the description being filled with words like 'jubilation', 'praise' and 'beauty', but is of such overpowering intensity that even the angels 'walk in terror' and 'faint with fear'. The interpenetration of these two themes is very typical of this literature. Verse 5 refers to the resurrection of the dead and final Judgement at the end of time. Note the magical *nomina barbara* in verse 6. It is clear that the pronunciation of these names, including theurgic utterance of God's unspeakable four-letter Name (verse 5), forms the ecstatic climax of the ascent. The mystic in the innermost Hekhal assumes the role of the High Priest in the Temple Sanctuary, since he and he alone was permitted to pronounce the proper Name of God. I hope that the translation conveys something of the power and resonance of these compositions, and one can perhaps begin to imagine their psychological effect, if recited in a state of high concentration and emotional intensity, probably following a period of ascetic preparation.

Many of these hymns include long lists of *nomina barbara* and a very large proportion include or end with Isaiah 6: 3 (the *Qedushah* or *Sanctus*). Indeed, Isaiah 6: 1–4, the vision and praise of the Divine Glory, is as central a text in this tradition as Ezekiel 1 itself:

> In the year that King Uzziah died, I saw the LORD sitting upon a high and exalted throne, and the train of His robe filled the Temple. Above Him stood the Seraphim. Each one had six wings, two covering his face, two covering his feet, and two with which he flew. And they were calling to one another: 'Holy! Holy! Holy is the LORD of Hosts! The whole earth is full of His Glory!' And the pillars of the door shook at the sound of their voices, and the House was filled with smoke.

It seems that the mystic, by combining recitation of these liturgical passages with visualisation of the images described, was able to enter, in imagination and belief, into the presence of the Glory and participate in the worship of the angels. As a result of his participation in the celestial liturgy, he assumes (like the High Priest in the Temple Sanctuary) a mediatory role between the heavenly and earthly worlds, bridging the gap between the community of Israel and its heavenly Father. Thus *Hekhalot Rabbati* 9: 2–3 (*Synopse* §§163–4):

> You are blessed by Me, and by heaven and earth, O descenders unto the *Merkabah*, if you tell and declare unto my sons what I do at the time of morning prayer, and at afternoon and evening prayer, each single time that Israel say before me: 'Holy! Holy! Holy is the LORD of Hosts!'
>
> Teach them, and say to them:
>   Raise your eyes towards the firmament, beyond your house of prayer, when you recite before Me: 'Holy! Holy! Holy is the LORD of Hosts!', for there is no dwelling place for Me in all the worlds that I have created. But at that time, when your eyes are raised towards My eyes, so are My eyes raised towards your eyes, at the time when you recite before Me: 'Holy! Holy! Holy is the LORD of Hosts!', for at that time the sound issues from your mouths, swirling and ascending before me like sweet incense.

And bear witness to them of that which you see of Me: what I do to the countenance of Jacob, their father, which is engraved by Me upon the Throne of My Glory. For at the time when you recite before Me: 'Holy! Holy! Holy . . .!', I bend down towards it and caress it, embrace it and kiss it, with my own hands and arms – three times, corresponding to the three times at which you recite the *Qedushah* before Me, as it is written: 'Holy! Holy! Holy is the LORD of Hosts!'

The form of the Glory is the subject of the *Shi'ur-Qomah* ('dimensions of the body') texts. Passages of this literature utilise the imagery of Isaiah 6: 1–4, 66: 1 and The Song of Songs to describe the enormous dimensions of the body of God as it appears to the visionary, seated upon the *Merkabah*. The following is a typical example (Cohen 1983: ll.47–58):

R. Ishmael said:

I saw the King of Kings, the Holy one, blessed be He, as He was sitting on an exalted throne, and His soldiers were standing before Him, to the right and to the left (cf. 1 Kings 22:19).

R. Ishmael says:

What is the measure of the body of the Holy One, Blessed be He, Who lives and endures for all eternity, may His Name be blessed and His Name be exalted?

The soles of His feet fill the entire universe, as it is stated in Scripture: 'The heavens are My seat, the earth, My footstool' (Isaiah 66:1).

The height of His sole is 30,000,000 *parasangs*: its name is *Parmeseh*.

From His feet to His ankles is 10,000, 500 *parasangs*. The name of His right ankle is *Atarqam* and the name of His left ankle is *Ava Tarqam*.

From His ankles until His knees is 190,000,000 *parasangs*: *Qanangi* is its name. The name of His right calf is *Qangi*; the name of the left is *Mehariah* . . . .

In the following passage of the text, we learn that these measurements are not in human *parasangs* (or Persian miles) but Divine ones, each being 120,000 times the length of the universe – so God is a very big chap! This may seem absurd, but Cohen is undoubtedly right to identify the basis of the text, which includes many hymns and prayers, as a theurgic liturgy intended to enable the mystic to experience the vision of the *kabod*. God's 'greatness' (*gedullah*) is punningly interpreted to mean (a) majesty,

(b) praise and (c) giant size. The underlying idea seems to be that the praise of the created universe, the angels, the community of Israel and the mystic himself is actually the 'substance' of the glorious form in which God manifests. In other words, the worship of Creation is what makes God visible and by 'magnifying' God the mystic causes Him to appear in his *kabod*. This is consistent with the picture found in *Hekhalot Rabbati*, ch. 3 (*Synopse* §§97–102), where it is said that God descends from the eighth heaven (where He dwells in Himself) to appear on the Throne of Glory in the innermost palace only at the time when the angels and Israel utter the *Qedushah*.

A central theme of these traditions is the transformation of the visionary (Morray-Jones 1992). There are numerous references in the Apocalypses, the *Hekhalot* writings and the midrashic traditions of the heavenly ascent to the metamorphosis of the mystic's body into a purified angelic or supra-angelic form of fire or light, which embodies or reflects the image of the Divine Glory and, like that Glory, expands to fill the universe. The classic instance of this is the transformation of Enoch into the supreme Archangel Metatron, who embodies the Image and the Name of God, described in the *Hekhalot*–Apocalypse 3 *Enoch* (tr. Alexander 1983):

When the Holy One, blessed be He, removed me from the generation of the Flood, He bore me up on the stormy wings of the Shekhinah to the highest heaven and brought me into the great palaces in the height of heaven . . . . (7: 1)

. . . I was enlarged and increased in size till I matched the world in length and breadth. (9: 2)

. . . the Holy One, blessed be He, fashioned for me a majestic robe, in which all kinds of luminaries were set, and He clothed me in it. He fashioned for me a glorious cloak in which brightness, brilliance, splendour, and lustre of every kind were fixed, and He wrapped me in it. He fashioned for me a kingly crown . . . and He called me 'The Lesser *YHWH*' in the presence of His whole household in the height, as it is written, 'My name is in him.' (12: 1–5)

When the Holy One, blessed be He, took me to serve the Throne of Glory, the wheels of the *Merkabah* and all the needs of the Shekhinah, at once my flesh turned to flame, my sinews to blazing fire, my bones to juniper coals, my eyelashes to lightning flashes, my eyeballs to fiery torches, the hairs of my head to hot flames, all my limbs to wings of burning fire, and the substance of my body to blazing fire. On my right –

> those who cleave flames of fire – on my left – burning brands – round about me swept wind, tempest and storm; and the roar of earthquake upon earthquake was before and behind me. (15.1f)

It seems that the vision of the Glory entailed the transformation of the mystic into the likeness of that Image of the Divine. This is frequently associated with the idea that the mystic 'assumes' or 'is clothed with' the Divine Name. This transformation was held to be extremely dangerous, should the mystic prove unworthy, but it seems to have been a central goal of the mystical endeavour. This motif is found in several Gnostic sources, and is surely the background of the language of 'glorification' found in the letters of Paul and other early Christian writings. The ascended and 'Glorified' Christ becomes an 'embodiment' of the Divine Image or Glory, and his followers participate in that spiritual–bodily transformation. Consider, for example:

> . . . those whom He foreknew He also predestined to be conformed to the Image of His Son, in order that he might be the first-born of many brethren . . . . (Romans 8: 29).

> He [Christ] will transform our body of humiliation in conformity with the body of his Glory by his power to subject all things to himself. (Philippians 3: 21)

> . . . and we all, beholding with unveiled faces the Glory of the Lord, are being changed into his likeness from one degree of glory to another . . . . (2 Corinthians 3: 18)

In the Epistle to the Hebrews, a text redolent with *Merkabah* imagery, Christ is represented as the transfigured and glorified High Priest in the celestial Sanctuary, who embodies both the Divine Image and the Name of God:

> . . . in these last days He has spoken to us by a Son, whom he appointed the heir of all things, through whom He created the world. He reflects the Glory of God and bears the very stamp of His nature, upholding the universe by His word of power. When he had made purification for sins, he sat down at the right hand of the Majesty on high, having become as much superior to the angels as the Name he has obtained is more excellent that theirs. (Hebrews 1: 2–4)

This is surely a Christian version of the ascending hero of the Apocalyptic and *Merkabah* tradition, who becomes mystically identified with the Divine Glory and so overcomes the gulf between the fallen, human world and the celestial realm of God.

Returning to the *Hekhalot* literature, the last three chapters of *Hekhalot Rabbati*, at the climax of the mystical ascent, contain a series of hymns which are said to be uttered by the Throne of Glory in the presence of God each day, and which the mystic himself is instructed to recite. This implies that he is identifying himself with the *Merkabah* and asking God to be enthroned upon or within him. In other words, he is seeking to become, like the patriarchs and righteous men of mythical history, a 'vehicle' for the manifestation of the Divine Image or Glory.

Curiously enough, the journey through the *Hekhalot* is referred to, sometimes as an 'ascent', but more frequently as a 'descent', and the mystics refer to themselves as *yordei–merkabah* or 'descenders to the Chariot'. The significance of this expression has never really been explained, but it may indicate that these mystics saw the journey as a descent within the body or the self, to the very centre of the soul (or the spiritual universe), where the Divine Image appears in all its Glory. A few sources state that the visionary adopts a foetal posture, with the head placed between the knees (the position adopted by Elijah on Mount Carmel, 1 Kings 18: 42), which perhaps supports this proposition (Scholem 1961: 49f). However, the journey also has an 'outer' dimension: the mystic is not only 'descending' within himself, but also 'ascending' to heaven. The structure of the seven *Hekhalot* is clearly based on the model of the Jerusalem Temple, with its concentric areas of increasing holiness (Maier 1964). The seventh *Hekhal* corresponds to the Holy of Holies where the Glory of God appears. The Temple was from an early period held to embody the structure of the Cosmos (Barker 1991), and the seven *Hekhalot* therefore correspond to the seven celestial levels. Some of the early Apocalypses employ a three-fold, rather than seven-fold, model and when Paul the Apostle states (1 Corinthians 12) that he was 'caught up to the third heaven' he is almost certainly referring to a visionary ascent/descent to the Innermost Sanctuary of the celestial or 'inner-worldly' Temple. Paul states that he does not know whether this experience occurred 'in the body' or 'out of the body'. It may be that the three structures: Cosmos, Temple and Body, were regarded as essentially the same reality, but on different 'scales'. The Epistle to the Ephesians, where body- and Temple-imagery are closely intertwined with the theme of ascent to heaven, becomes very interesting if read in the light of these ideas: there, the Church (and each individual member thereof) is the 'Resurrection Body' of the exalted and Glorified Christ, and the Temple of the Indwelling God (2: 19–22). Consider, in the light of the traditions that we have been discussing, Ephesians 4: 7–13:

> There is one Body and one Spirit, just as you were called to the one hope of your calling: one Lord, one faith, one

baptism, one God and Father of all, who is above all and through all and in all.

But to each one of us was given grace according to the measure of the gift of Christ. Therefore it is said: 'When he ascended on high, he made a captive of captivity: he gave his gifts to men.' (Psalm 68: 18)

When it says: 'he ascended', what does it mean but that he also descended into the lower parts of the earth? The one who descended is the one who ascended far above all the heavens, that he might fill all things.

And he gave some to be apostles, and some to be prophets, and some to be evangelists, and some to be shepherds, and some to be teachers, for the equipping of the holy ones for the work of ministry, for the building up of the body of Christ, until we all arrive at the unity of the faith and full knowledge of the Son of God, at a man of complete maturity, at the measure of the stature of the fullness of Christ.

Here, Christ is identified with the Divine Glory that fills the universe (cf. Isaiah 6: 3). The members of his Church participate in that spiritual–bodily transformation, which is described in terms that are highly reminiscent of the *Shi'ur Qomah* ('. . . the measure of the stature of the fullness . . .'). Christ, his Church and its individual 'members' have become the Body of God's Glory.

REFERENCES

Alexander, Philip S. (trans.) (1983). 3 (Hebrew Apocalypse of) Enoch. In Charlesworth, James H. (ed.) *The Old Testament Pseudepigrapha*, vol. 1. London: Darton, Longman and Todd.

Barker, Margaret (1991). *The Gate of Heaven: The History and Symbolism of the Temple in Jerusalem.* London: SPCK.

Cohen, Martin S. (1983). *The Shi'ur Qomah: Liturgy and Theurgy in pre-Kabbalistic Jewish Mysticism.* Lanham, New York and London: University Press of America (includes a translation of *Sepher Ha-Qomah*).

Gruenwald, Ithamar (1980). *Apocalyptic and Merkavah Mysticism.* Leiden: Brill.

—— (1988). *From Apocalypticism to Gnosticism.* Frankfurt am Main, Bern, New York and Paris: Peter Lang.

Halperin, David J. (1980). *The Merkabah in Rabbinic Literature.* New Haven, Connecticut: American Oriental Society.

—— (1988). *The Faces of the Chariot: Early Jewish Responses to Ezekiel's Vision.* Tübingen: Mohr/Siebeck.

Maier, Johann (1964). *Vom Kultus zur Gnosis*. Salzburg: Otto Müller.

Morray-Jones, C. R. A. (1988). *Merkabah Mysticism and Talmudic Tradition*. University of Cambridge PhD dissertation.

—— (1992). Transformational Mysticism in the Apocalyptic–Merkabah Tradition. In *Journal of Jewish Studies*, 43: 1–31.

Newsom, Carol (1985). *Songs of the Sabbath Sacrifice: A Critical Edition*. Atlanta: Scholars Press.

Rowland, Christopher C. (1982). *The Open Heaven: A Study of Apocalyptic in Judaism and Early Christianity*. London: SPCK.

Schäfer, Peter (1984). Merkavah Mysticism and Rabbinic Judaism. *Journal of the American Oriental Society*, 104: 537–44.

—— (1986). *Gershom Scholem Reconsidered: The Aim and Purpose of Early Jewish Mysticism*. 12th Sacks Lecture, Oxford Centre for Postgraduate Hebrew Studies. [also in Schäfer (1988) *Hekhalot–Studien*. Tübingen: Mohr/Siebeck.]

Scholem, Gershom G. (1961). *Major Trends in Jewish Mysticism*. New York: Schocken (1st edition, 1941).

—— (1965). *Jewish Gnosticism, Merkabah Mysticism and Talmudic Tradition*. New York: Jewish Theological Seminary of America (1st edition, 1960).

Synopse: Schäfer, Peter (ed.) (1985–91). *Synopse zur Hekhalot–Literatur* and *Übersetzung der Hekhalot–Literatur*. Tübingen: Mohr/Siebeck.

WENDY PULLAN

Mapping Time and Salvation: Early Christian Pilgrimage to Jerusalem

The overall urban form of Jerusalem as a sacred city has always been difficult to discern: it does not have a sacred geometry. There are instead overlays of the three major monotheistic traditions in cycles of construction and destruction over more than three thousand years, and many of the holy places are shared. The physical unity and identity of the city is outwardly more apparent in its materials, light and relation to natural topography. Nonetheless, conceptions of space on an urban scale, which are unique and specific to each of the three religions, are manifested in various ways.

As far as the Christian city is concerned, it was characterised from the beginning by an extreme ambiguity between earthly and heavenly cities. During the first three centuries of Christianity, material representation of any sort, in any place, was minimal; when an ecclesiastical building programme was initiated in the fourth century by the Byzantine emperor Constantine, it was essentially a series of commemorative churches and shrines. However, it is my opinion that far from being an unrelated and loose collection of sacred sites, the notion of Jerusalem as a holy city, symbolised by architectural and urban form, was incipient as early as the latter part of the fourth century. Certainly it was during the early Christian and Byzantine period, which lasts from the proclamation of Jerusalem's honoured status in 325 at the Council of Nicaea[1] until the Muslim conquest in 638, that there evolved a well unified sense of a Christian holy city. This understanding was based on the establishment of a ritual topography that was related to the fourth century development of the Christian stational liturgy founded on the holy sites and the Scriptural events associated with them.

In the same century pilgrimage to the Holy Land became

prevalent,[2] and while we cannot speak of pilgrimage in a sacramental sense or even as a theology, it is a phenomenon which was closely linked to the development of ritual topography, and, in its own participatory characteristics, can be extremely revealing in terms of what meaning the sacred space of the city had at the time. Early Christian pilgrimage to the Holy Land was an embodiment of an itinerary of the soul and its salvation. In such a journey, or *peregrinatio*, pilgrims mediated the space between two worlds, the Earthly and Heavenly Jerusalems, and as they made their way they explored and mapped a landscape of both the biblical past and the eschatological future. Their territory is one of time interpreted by space, the temporality of the Earthly Jerusalem as it prefigures and aspires to the eternity of the Heavenly. This chapter attempts to investigate this territory and its mediative nature as manifested in pilgrimage.

Human representation – whether literature, art, architecture, or liturgy and pilgrimage – attempts to mediate the space between sacred and profane (Gadamer 1979: ch. 1.II.2, esp. 132–3). In Christianity, God's word is primarily and most explicitly revealed through Christ as Logos and transmitted to humanity in Scripture. In the Logos 'the Word became flesh, and dwelt among us' (John 1: 14). Likewise, the traditional Hellenistic and Jewish notions of Temple and Holy City were also interpreted as spiritually residing in Christ (John 4: 21–4; Matthew 18: 20). However, the Christian city did eventually become embodied. As a reflection of the heavenly, the earthly city is less articulate and explicit than Scripture, but it is a more concrete form of representation.[3]

In Jerusalem such representation began in 326 with three Constantinian basilicas built on three sites: the Church of the Nativity to the south of Jerusalem in neighbouring Bethlehem[4] on the site of the birthplace of Jesus; the Eleona Church to the east of the city on the upper slope of the Mount of Olives, marking the place where Jesus taught the Mysteries to his disciples; and the Martyrium basilica and Anastasis rotunda, today known as the Church of the Holy Sepulchre, on the Hill of Golgotha in central Jerusalem on the site of the crucifixion and resurrection. It is significant that three holy places were chosen, breaking with the Jewish tradition of the Temple as a single, unique holy place. By the end of the fourth century, several more churches were added and the pattern of ritual topography had coalesced: major churches on three hilltops[5] – Golgotha in the centre, Mount Sion[6] to the south, and Mount of Olives to the east; a church beyond Mount Sion about eight kilometres to the south at Bethlehem and

Mapping Time and Salvation

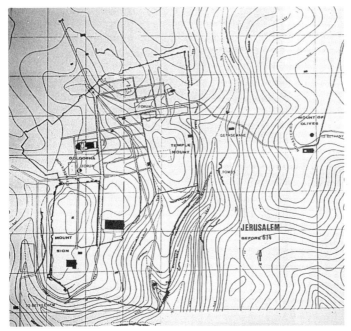

Figure 1   Map of Byzantine Jerusalem before the Persian invasion of 614.

one beyond Mount of Olives about five kilometres to the east at Bethany completed the pattern. Saint Jerome, who lived in Bethlehem from 386 until his death in 420, produced a map of the Holy Land which is no longer extant, but a twelfth century copy (Nebenzahl 1986: 19) shows Jerusalem as a circular walled city with these features of the landscape to the south and east.

The Persians destroyed much of Christian Jerusalem in 614, and today no buildings from the period are extant. But until the Persian invasion, churches were continually added to the landscape, in most cases on sites deemed to be holy by their connection to events in the passion and resurrection of Christ, or occasionally Old Testament events (Figure 1). The majority of these sites were established on or near the eastern and southern routes from Golgotha, their locations determined by a variety of means including common opinion and oral histories, the relocating of earlier New and Old Testament sites to a somewhat revised Christian configuration, and the discovery of relics often based on dreams and visions; the process can be characterised as a refining of the ritual topography within an expanding mythology.[7] Jerusalem became a city filled with holy places, yet far from fragmenting the city, this tendency appears to have unified it because of each site's revelatory status and place within the holy narrative. Described by Cyril, Bishop of Jerusalem from *c.* 349 until 386:

> It is particularly fitting that as we speak of Christ and Golgotha here on Golgotha, so also we should speak of the Holy Spirit in the Upper Church [on Mount Sion]; but since he who descended there shares the glory of Him who was crucified here, therefore we speak here also of Him who descended there, for their worship is indivisible. (Cyril 1970a: 16.4)

Continuity of place and time was further accomplished by Cyril's development of the Jerusalem liturgy (Dix 1986: 349–50). Catechetical lectures and early lectionaries[8] attest to the importance of the holy sites in the Jerusalem liturgy, but it is the account of Egeria, a pilgrim to the Holy Land from 381–4 when Cyril was bishop of Jerusalem, which gives some of the most useful and vivid descriptions of the liturgy as it operated within the city. The Jerusalem liturgy is a stational liturgy, moving from church to church and site to site. Scriptural readings and prayers relating to the significance of a particular site were recited at that site, at the appropriate time of day, week or year. For example, on each Sunday the cave of the resurrection was filled with the smoke of incense and the bishop read the account of the resurrection at the opening of the cave (Egeria 1981: 24.10). But the Sunday Pentecost services, commemorating the descent of the Holy Spirit, were held at the basilica on Mount Sion at the same place where the event was believed originally to have taken place (Egeria 1981: 25.6). Easter week was spent moving from site to site: on Thursday, according to Egeria, to 'the church on Eleona which on this very day the Lord visited with the apostles' (35.2–3); and then the night at Gethsemane for 'a reading about the Lord's arrest' (36.3); Friday before dawn through the city to the hill of the crucifixion with a 'passage about the Lord being led away to Pilate' (36.3–4); and before the sun is up, 'those with the energy then go to Sion to pray at the column at which the Lord was scourged' (37.1); and so on. Egeria remarked a number of times on such appositeness.

> And what I admire and value most is that all the hymns and antiphons and readings they have, and all the prayers the bishop says, are always relevant to the day which is being observed and to the place in which they are used. They never fail to be appropriate. (47.5)

Mapping Time and Salvation

Processions, often led by the bishop, linked the sites; usually they started, ended, or in some way focused upon the Anastasis and Martyrium on Golgotha. From there, the liturgy spilled out into the streets and hills and valleys; it filled the city with songs and lamentations, and with the flickering lights of candles and the smell of incense. It even entered people's homes during feast days where eating and sleeping were part of the general schedule (Egeria 1981: 35–6). It was truly a liturgy of the city; to worship in late fourth century Jerusalem was to be constantly and exhaustingly on the move.

Thus, the ritual topography of early Christian Jerusalem may be described as centring on its most important site, Golgotha, which is joined by processional routes, mostly to the east and south, to many other sacred sites. The movement from site to site links them and allows the continuity of Scriptural narrative; because of movement the city is given further unity and comprehensiveness. Ultimately, Christian unity derives from the person/God of Christ; in him the idea of movement is emphasised by his statement: 'I am the way, and the truth, and the life; no one comes to the Father, but by me' (John 14: 6). In these words are fused orientation, passage and centre, as well as the spectrum of time from history to eternity. It was this world in which the pilgrim to Jerusalem participated.

Early pilgrims to the Holy Land partook of the Jerusalem Liturgy, yet as *peregrini*, or foreigners,[9] they also had their own agendas and rituals at the holy sites (Egeria 1981: 10.7). Gerhard Van der Leeuw wrote that in its broadest sense, any visit to a holy place may be called a pilgrimage (1986: 401). In early Christian Jerusalem such a wide understanding is necessary in considering who might be called a pilgrim: for example, Saints Jerome and Paula left Rome to live spiritual existences in Bethlehem, in a sense making pilgrimages of their whole lives. It is also worthwhile to acknowledge the scope which derives from the delicate balance between mental and physical pilgrimage: Sophronius, bishop of Jerusalem from 634–8, whose *Anacreontica* poems about walking through Jerusalem (Sophronius 1977), likely composed during a period of exile, can be regarded as a pilgrimage in memory and anticipation. But despite the advantages of a broad understanding, Van der Leeuw's definition is lacking for it does not emphasise the importance of the journey. Far from being simply a visit to a holy place, the idea of movement, a prime factor in the stational liturgy, is to an even greater extent the essence and determining factor of pilgrimage.

From site to site throughout the Holy Land, the pilgrim's

itinerary was within a continuum between sacred and profane; sometimes the journey included a more holy and sometimes a less holy place. Many of the sites themselves were defined by architecture, or at times only by local customs and claims to give a sort of mental boundary. These sites were not separated from the rest of the world in the Eliadian sense of opposition of sacred and profane (Eliade 1959: 10), but part of an extraordinary and charmed world in which pilgrims travelled. Accounts reveal dedication, captivation and even ecstasy upon the part of the *peregrini*; Jerome described Paula as 'visiting the holy places [where] so great was the passion and enthusiasm she exhibited for each, that she could never have torn herself away from one had she not been eager to visit the rest' (Jerome 1893: 108.9). Sophronius' poems are full of dynamic joy and yearning as he imagined himself moving through the city:

> A divine longing for holy Solyma
> presses upon me insistently.
> Let me walk thy pavements . . .
>
> Exultant let me go on to the place
> where all of us
> who belong to the people of God
> venerate the glorious Wood of the Cross . . .
>
> And, speeding on,
> may I pass to Sion
> where, in the likeness of fiery tongues,
> the Grace of God descended . . .
>
> And soon may I come, consumed
> with the heat of a holy desire
> to the townlet of Bethlehem,
> birthplace of the King of All.
> (Sophronius 1977: 19.5–7, 35–8, 55–8, 20.23–6)

A pilgrim's progress was dependent upon both a literal and symbolic understanding of Scripture and physical place. However, the meaning of pilgrimage appears to be not just a duality of embodiment and articulation, but to encompass a number of senses or levels of meaning. These may be described as: re-enactment, witness, salvation and eschatology.[10]

Re-enactment and witness are closely related; both are functions of time. Re-enactment works to bring paradigmatic time into the present while witness attests to the sacred events and, in doing so, is a confirmation of truth in the present. Cyril of Jerusalem saw as witnesses to the events the sites themselves (1970a: 14.22–3), in addition to human witnesses who, by seeing the places, could testify to the faith (1970a: 13.23). According to Paulinus, the early fifth century Bishop of Nola, pilgrims visited the sites in order to 'see and touch the places where Christ was physically present' (1967: 49.14). But what they were really after was truth: like the inhabitants of Jerusalem, to 'touch as it were with their own hands the truth each day through the venerable places' (Cyril of Scythopolis 1991: 154.15). The Christian idea of witness, translated from the Greek *martus*, or martyr in English, was originally one who had seen the resurrected Christ (Luke 24: 46–8; Acts 1: 21–2), i.e. one who could bear witness, and in the later persecutions, bear witness to the faith even until death. After the early fourth century, martyrs were no longer made by pagan Rome, but the ideal of suffering with Christ remained: 'As Scripture says, if we suffer with Him we shall also reign with him' (Paulinus 1967: 31.2), and pilgrimage, as a long, arduous and often dangerous act, was regarded as an imitation of such suffering which might help to lead the pilgrim to Heaven (Jerome 1893: 46.13; Valerius 1981: 4a).

Reading the appropriate Scriptural passage at the site enhanced the meaning of both text and place, but sometimes re-enactment was more physical: Egeria followed the route of the children of Israel in the Sinai Desert (1981: 1.1ff.); the sixth century Piacenza Pilgrim threw stones at the place where David met Goliath (Piacenza Pilgrim 1977: 31). At times, re-enactment had a visionary element: Saint Jerome stated that in Bethlehem Paula 'protested in my hearing that she could behold with the eyes of faith the infant Lord wrapped in swaddling clothes and crying in the manger' (1893: 108.10). Emperor Constantine's mother Helena, who visited the Holy land in 326, was said to have located the cross of the crucifixion buried deep in the ground of Golgotha through a vision, and was then able to testify to its authenticity by resurrecting a dead person through its touch (Paulinus 1967: 31.5–6).[11] The visions of both Paula and Helena coupled past and present time, but Helena's went even further to reinforce a past tradition by recreating it in a present context on the same site; in the sense that her act is both imitative and creative it increases in its ontological reality (Gadamer 1979: 124).

In all these deeds of witness and re-enactment the pilgrim focused on past events; but salvation, the third sense of pilgrimage, is future oriented. This had to do particularly with the salvation of the soul. Augustine reflected on this general preoccupation in his many attempts to describe and interpret the progressive stages which human beings must pass through in order to find holiness (Augustine 1950: 33.70–6; 1961: 7.17); this is *peregrinatio animae*, pilgrimage of the soul. Pilgrimage to Jerusalem, although a physical act, was very much a matter of the soul's salvation. A seventh century Galician monk called Valerius puts this clearly in a letter written to his brother monks in praise of Egeria, who after 300 years was still seen as a model for the holy life. Valerius wrote that in seeking salvation, Egeria's 'heart and her whole being were on fire with an earnest longing to seek the kingdom of heaven on high' (1981: 3). He believes she accomplished this progressively, 'for the more she had advanced in holy doctrine, the more insatiably her holy longing burned in her heart' (1b). Egeria's capacity for holiness and her longing for it are reciprocal, but so seems to be the relationship between the pilgrim's physical journey and the soul's journey. The physical journey could become a disposing of physicality; Paulinus described his friend Melania as:

> Abandoning worldly life and her own country, she chose to bestow her spiritual gift at Jerusalem and to dwell there in pilgrimage from her body. She became an exile from her fellow citizens, but a citizen amongst the saints. With wisdom and sanctity she chose to be a servant in this world of thrall so as to be able to reign in the world of freedom. (1967: 29.10)

The fourth sense of pilgrimage is eschatological, which is closely connected to its soteriological meaning. The early Christian hope for salvation works on two levels: a personal one to which an individual can partially progress in this life, and one that involves all of humanity at the end of time. The individual is in the end connected to the whole human race in the reality of last judgement; the ultimate redemption and final destination of the soul's journey is eternity. This is soteriology become eschatology, and it must be seen in terms of time rather than space. We have already seen the element of time at work in pilgrimage's sense of witness and re-enactment, but this was essentially in terms of the past, whereas eschatology is hope for the future.

Valerius wrote of Egeria: 'Here, by her own will and choice she accepted the labours of pilgrimage, that she might, in the choir of holy virgins with the glorious queen of heaven, Mary the Lord's mother, inherit a heavenly kingdom' (1981: 4a). By pilgrimage, heaven might be attainable in eternity, but pilgrimage to the Holy City also had a more concrete meaning. Jerusalem plays a major role in the eschatological future both as the location of final judgement (Cyril 1970a: 14.16) and in its prefiguring of the Heavenly Jerusalem. The movement of the pilgrim through the city was in itself a mapping of what will be met in heaven at the end of time; this is put explicitly by Valerius:

> she [Egeria] will return to that very place where in this life she walked as a pilgrim, and there with the holy virgins and saints she will meet the Lord in the air at his coming. Her lamp will be alight with the oil of bright holiness . . . . (1981: 4b)

We might ask if the eschatological role of Jerusalem is an abstract idea, or are more specific meanings attached to the physical city and identified with places in it? To explore this, it is necessary to return to the topography, both natural and ritual, as it was understood and interpreted in the early Christian period.

Jerusalem sits on a geographical divide between a relatively green and fertile Mediterranean landscape and climate to the west and the desert to the east (Kutcher 1973: 11). This edge is quite dramatic: from the peak of the Mount of Olives the direction of the city is green but the other way to the desert is brown. Mount Sion, at the south-west corner of the city, is fertile and green. Its crest is in the walled city and it juts out beyond the wall to an area which was enclosed by a wall built by the Byzantine Empress Eudocia in about 450. By its physical position, Mount Sion is very much part of the city, either densely built or cultivated through much of history. Eudocia's wall (which no longer exists) indicated that the early Christians perceived it as such, as does the construction of a wide classical street, the Cardo, from the area near Golgotha out to Mount Sion.[12] The Mount of Olives, on the other hand, was not uniformily considered part of Jerusalem in the fourth century (Walker 1990: 217ff). It is separated from the city by the Valley of Jehosephat (today usually known as the Kidron valley), and has always been reached by a narrow, twisting road going into the valley and up its slope. The Mount of Olives stands open on either side between Jerusalem and the Judaean Desert; from the fourth century its peak was a favoured place for monasteries which were perched overlooking the history of the city and the eternity of the wilderness. At slightly over 800 metres, it is the highest point in the city. These two hills, Sion and Olivet, which along with Golgotha figure so strongly in the ritual topography of early Christian Jerusalem, are clearly contrasted in an urban and geographical sense. Mount Sion is earthly, associated with the vitality of the urban centre; Mount of Olives is chthonic and celestial, detached from the town and its life.

The Mount of Olives, as a place, is associated largely with death, ascension, the eschaton and eternity; it could be referred to as the 'eschatological' direction or route from Golgotha. We have seen how this corresponds to its natural topography; its religious traditions also reflect these meanings. As the hill top where Jesus is believed to have ascended to heaven, it is a true Eliadian *axis mundi*; Paulinus wrote:

> He was taken into a cloud . . . . That single place and no other is said to have been so hallowed with God's footsteps that it has always rejected a covering of marble or paving. The soil throws off in contempt whatever the human hand tries to set there . . . [and] preserves the adored imprint of the divine feet in that dust trodden by God, so that one can truly say: we have adored in the place where his feet stood. (1967: 31.4)

In the late fourth century the Church of the Ascension, a centrally planned building capped by a 'dome of heaven', was built over the imprinted ground. The concern with death and eternity extends to Bethany, on the eastern, desert side of the Mount of Olives, where a church was built to commemorate the raising from the dead of Lazarus. On the western side of Olivet is the Valley of Jehosephat, from the Hebrew 'God judges' (Joel 3.2,12). It contains Jewish cave tombs, including that of Zechariah which was adopted by subsequent Christian generations as the burial place of James, the first bishop of Jerusalem (Theodosius 1977: 9). Particularly important for fourth century Christians were eschatological prophecies for the Mount of Olives: Christ ascended from Olivet and the Second Coming is also supposed to come from the east, this having been prefigured in the past by Jesus' entry to Jerusalem from the Mount of Olives on what is today known as Palm Sunday, and by post-ascension visions such as a luminescent cross suspended in the sky from Golgotha to Olivet (*Codex Armenien* 1971: 54; Cyril 1970b: 4, 6). In anticipation of the world to come, Jesus was said to have taught his disciples the eternal Mysteries in a cave just below the peak of Olivet, this marked in the fourth century by the Eleona Church (Sophronius 1977: 19.1–12; Eusebius 1920: 6.18.288). In general, Christ is symbolised by the easterly direction and the rising sun (Danielou 1960: 31–4); medieval maps and churches are oriented to the east, and liturgical movement in church is from west to east. The parousia will be from the eastern sky (Cyril 1970a: 14.30); in the meantime, we as humans on earth face towards the east and make our way towards its light, in hope and anticipation.

Mount Sion is associated with traditions which are very much part of the living city and temporal history, particularly in terms of the incarnation and the prefiguring and founding of Christianity as a religion for humanity on earth. It could be called the 'incarnational' axis or direction. Paulinus combined two passages from Psalms to describe Mount Sion: 'May the Lord bless you out of

Sion, and may you see the good things of Jerusalem all the days of your lives, so that our portion may be jointly in the land of the living' (1967: 40.12). By the fourth century, certain formulative aspects of the new religion, such as the descending of the Holy Spirit at Pentecost, were said to have occurred on Mount Sion, and by the fifth century the Armenian Lectionary (*Codex Armenien* 1971: 39) associates it with the Last Supper and the giving of the first eucharist. These two traditions – mediative action of Spirit and liturgical sacrament – aided in making Christianity operative for human practice. Considerable fourth and fifth century interest in the origins of Christianity enhanced certain Mount Sion traditions. Because of associations with primitive Christianity, Holy Sion Basilica was called the 'Mother of all Churches' (Cyril of Scythopolis 1991: 153) and the throne of Saint James was placed there.

Feast days for James (Jerusalem's first bishop), Stephen (the first Christian martyr) and even King David (both James and David were considered to have been relatives of Jesus) were held in the basilica (Limor 1988: 491–3). Also indicative of the concern with origins are a number of churches devoted to Mary. The route south from Mount Sion leads to Bethlehem and the Church of the Nativity; along the way is the mid-fifth century Kathisma Church on the spot where Mary was said to have stopped to rest while in labour. In 543 the huge Nea Church of Maria Theotokos,[13] i.e. the New Church of Mary the Mother of God, was built by the Emperor Justinian just north of Holy Sion Basilica. There is a tradition that Mary spent the rest of her life residing on Mount Sion, and writers spoke of Mary 'falling asleep' there (Wilkinson 1977: 172), and even her death is combined with her ability to give life: 'Rock [Sion] where Mary, handmaid of God, childbearing for all men, was laid out in death' (Sophronius 1977: 20.64–6). But, perhaps significantly, her tomb (as well as that of Saint James) is venerated on the eastern, eschatological side of the city in the Valley of Jehosephat.[14]

The third Christian hilltop, Golgotha, is gently sloped within the walled city.[15] The early Christians saw it as the new Holy of Holies (Eusebius 1890: 3.33),[16] the centre of the world (Cyril 1970a: 13.28); certainly, it was depicted as the centre of Jerusalem in the sixth century mosaic Madaba Map. It is where processional routes converged, in a sense the pivot point between the incarnational axis from Sion and the eschatological route from Olivet. As the site of both crucifixion and resurrection, it is indeed the point where historical time meets transcendent time, the centre of sacred space and time. Cyril stated that on this hill, because of

Christ's burial, 'peace was made between heaven and earth' (1970a: 14.3). Eusebius called it the New Jerusalem (1890: 3.33). Over the fourth to seventh centuries, ancient holy places, such as the site of the creation of Adam (*Brevarius* 1977: 2) and that of the sacrifice of Isaac (Theodosius 1977: 4) were 'relocated' to Golgotha, seemingly an attempt to reach back to the primordial, to unite historical events of the distant past with the transcendent future. Golgotha is the place where affirmation of the historical process, prefigured in Judaism and realised in the incarnation and crucifixion of Christ, leads to transcendence through redemption in Christ's resurrection. Golgotha was in Eusebius' words 'a monument to the Saviour's victory over death' (1890: 3.33; see also Eusebius 1976: 9.16).

Many pilgrims' accounts make it sound as if they accomplished all the sites of Jerusalem in a day's time. In fact, most pilgrims spent far longer than one day; Egeria made the city her base for three years (1981: 17.1). It is unlikely that they followed a neat pattern of movement between Sion, Olivet and Golgotha along the routes of sacred space and time. In addition to this, there were other sites in the city which did not fall so exactly into the pattern of incarnational and eschatological routes.[17] Nonetheless, the pilgrims would have participated in liturgical processions which followed very distinct routes based on Scriptural narrative, and the traditions associated with the three hills and their churches were well known.[18] As early as 335, Eusebius described the original three places of theophany: Bethlehem, Olivet and 'between these [Golgotha] at the scene of the great struggle, signs of salvation and victory' (1976: 9.17). The unifying sense of time is attested to by Cyril: 'For He descended from heaven to Bethlehem, but from Mount Olivet He ascended to heaven; He began in Bethlehem His struggles for men, but on Olivet He was crowned for them' (1970a: 14.23). The early Christians recognised that churches may have been built by emperors, but as Paulinus noted, the city was God's plan (1967: 31.4), and God is aware of all that is hidden there (31.5).

We have seen that this plan, revealed in Scripture, was embodied in the sites of the city; the narrative became comprehensible through the linking of the sites by processional routes and the movement between them. But in movement, not only historical narrative was related: in addition to this, the pilgrim mapped the journey of the salvation of his soul through temporal history to eternity. On earth, such cartography could only be a prefiguration, a symbol, but, in such action, the hope for the eschaton was made possible. The key was movement, backwards through history to

Figure 2 Jerusalem in the Madaba Map.

the founding holy events and forwards to the end of time. This is attested to not only by the stational liturgy and the pilgrims' accounts but by the very revealing visual representation of the Madaba Map (Figure 2). This mosaic, laid in the floor of a church in Madaba, Jordan, is a depiction of the sacred topography in Palestine with Jerusalem in the centre (Avi-Yonah 1954: 10–16). Its meaning still leaves a lot to be deciphered and is far beyond the scope of this chapter; but one overwhelming aspect of the map should be noted here. In the vignette of Jerusalem, the Anastasis and Martyrium on Golgotha are shown in the centre of the city although this was not geographically accurate. However, the real centre piece is not the church but the road in front of it, the Cardo. Archaeological excavations have demonstrated that with a width of 22 metres and a colonnade on either side it would have been a monumental street. It has been depicted in the Madaba Map in such a manner that the street takes over Jerusalem; it shows a city perceived to be always 'on the way'.

NOTES

I am grateful to Peter Carl for his comments on this chapter.

1 Seventh Canon, Council of Nicaea, cited in Peters 1985: 131.
2 The first known Christian pilgrim to the Holy Land was Melito of Sardis in c. 160. There are records of a few others before the fourth century, but it was only after the Council of Nicaea that pilgrimage developed into a frequent and important practice.
3 See Vesely 1987: 31–5 for a discussion of the relationship between architecture and language within poetic mythos.
4 Bethlehem, although a separate town, was considered within the area of Jerusalem's ritual topography; see Cyril 1970a: 16.4.
5 A fourth hilltop, Mount Moriah, was the site of the destroyed Jewish Temple. Its desolation demonstrated, in the words of church historian Eusebius of Caesarea, 'the effect of Divine judgement on its impious people' (1890: 3.33).
6 Mount Sion or Zion has a long, complicated history, with three locations and two spellings. When transliterated from the Hebrew, it is spelled 'Zion' and when transliterated from the Greek, it is 'Sion'. As far as location is concerned, both the Davidic city and the Temple have been identified with Zion; these were two hills in the eastern portion of the city, today known as David's City and the Temple Mount. It is uncertain why Mount Sion of Christian tradition is on yet a third hill, the south-western hill. In the Byzantine period Mount Sion referred to both the area inside the city wall (today's Armenian Quarter) as well as what is today called Mount Zion on the southern portion of the hill outside the city wall.
7 Refining the ritual topography became a compacting of it as well; by the thirteenth century, many of the sites had been relocated along a portion of the eastern processional route now known as the Via Dolorosa.
8 For example: Cyril 1969, 1970; *Codex Armenien* 1969, 1971.
9 In classical Latin, *peregrinus* means 'of foreigners or foreign places, foreign' and 'of animals: a roaming, ranging' (*Chambers Murray Latin–English Dictionary*. London and Edinburgh, 1989: 524). In Early Christianity life on earth was seen as an exile from the Heavenly Jerusalem; the Greek *paroikia*, meaning 'the stay or sojourn of one who is not a citizen in a strange place', became 'parish . . . a community of strangers' (*A Greek–English Lexicon of the New Testament and Other Early*

Christian Literature. Chicago University Press, 1979: 629), the basic unit of the Christian Church. The Latin *peregrini*, which implied wandering as well as exile, was used by the Christians: in Augustine, the soul has wandered from its true home and must return.

10 Many of the Church fathers, including Origen, Jerome and Augustine, wrote about the literal and symbolic senses of Scripture. John Cassian (*c*. 360–435) described four senses: historical, allegorical, tropological or moral, and anagogical, which he then applied to the city of Jerusalem. Historically, he said, Jerusalem is the city of the Jews; allegorically, the city of Christ; tropologically, the soul of man; and anagogically, the heavenly city of God, mother of us all (Cassian 1894: 14.8). Likewise, a pilgrim's understanding and interpretation of the holy places must be seen on various levels of meaning, and while there is not always a direct correspondence, it is useful to remember Cassian's four senses when looking at the pilgrim's symbolic world.

11 The story of the discovery of the cross was not fully developed until around the end of the fourth century (Wilkinson 1977: 175.3). Eusebius does not mention it; if there really was such a discovery, and whether Helena was involved, is uncertain. Nonetheless, the story, and Helena's role in it, quickly became a popular and important tradition as shown in Paulinus' narrative.

12 The Cardo was aligned by the Romans before AD 325, and they appear to have constructed its northern portion. But the southern extension from the centre of Jerusalem to Mount Sion was built in the Byzantine period, probably in the sixth century. For a summary of the archaeological excavations, see Avigad 1983: 211ff.

13 The intention of the builders of the Nea Church, the largest Byzantine church in Palestine, is somewhat mysterious. It was one of a number of Theotokos churches constructed in the Byzantine Empire after the Councils of Ephesus (431) and Chalcedon (451) accepted the orthodoxy of Mary as mother of the son of God. The Nea Church has always been enigmatic because in Jerusalem it alone is located on a site which has no known holy tradition, and yet the difficulty of construction on the steeply sloped hillside would indicate that the building plot was chosen for a particular reason. If the church was intended as part of a Mary cult on Mount Sion, that would help to explain its position.

14 Mary's tomb is located in Gethsemane towards the north end of the Valley of Jehosephat.

15 Wilkinson (1977: 177) notes that Golgotha 'was in no sense a hill or "mount".' Nonetheless, it is important to note that people perceived it to be a high place. Cyril described it as 'conspicuous in its elevation' (Cyril 1969: 10.19).
16 The Jews had understood the Holy of Holies in the Temple to be the centre of the world.
17 The two most significant sites which exist somewhat outside of the routes are the Pool of Siloam to the east of Mount Sion, and the Church of St Stephen north of the city wall.
18 Many of the traditions and mention of them in written sources have been compiled in the gazetteer in Wilkinson 1977: 148–78.

REFERENCES

Primary Sources (Citations in the text are according to chapter or verse number.)

Augustine (1950). *On the Greatness of the Soul*. In Collerhan, J. M. (trans.) *Ancient Christian Writers* 9, pp. 1–112, 189–220. New York and Ramsey, NJ: Newman Press.
—— (1961). *Confessions*. Pine-Coffin, R. S. (trans.). Harmondsworth, Middlesex: Penguin.
Brevarius (Or Short Account) of Jerusalem (1977). In Wilkinson, John (trans.) *Jerusalem Pilgrims Before the Crusades*, pp. 59–61. Jerusalem: Ariel and Warminster: Aris and Phillips.
Cassian, John (1894). *The Conferences*. In Gibson, Edgar C. S. (trans.) *Nicene and Post-Nicene Fathers*, 2/11, pp. 295–545. Oxford and New York.
Le Codex Armenien Jerusalem (1969, 1971). Renoux, Athanase (trans.) *Patrologia Orientalis* 163, 168, 2 vols. Tournout.
Cyril of Jerusalem (1969, 1970a). *The Lenten Lectures (Catecheses)*. In McCauley, Leo P. and Stephenson, Anthony A. (trans.) *The Works of Saint Cyril of Jerusalem*, 2 vols. Washington DC: Catholic University of America Press.
—— (1970b). *Letter to Constantius*. In McCauley, Leo P. and Stephenson, Anthony A. (trans.) *The Works of Saint Cyril of Jerusalem*, vol. 2, pp. 231–5. Washington DC: Catholic University of America Press.
Cyril of Scythopolis (1991). *Life of Sabas*. In Price, R. M. (trans.) *Lives of the Monks of Palestine*, pp. 93–219. Kalamazoo, Michigan: Cistercian Publications.
Egeria (1981). *Travels*. In Wilkinson, John (trans.) *Egeria's*

Travels to the Holy Land, pp. 89–147. Jerusalem: Ariel and Warminster: Aris and Phillips.
Eusebius of Caesarea (1890). *The Life of Constantine*. In Richardson, E. C. (trans.) *Nicene and Post-Nicene Fathers*, 2/1, pp. 481–559. Oxford and New York.
—— (1976). *In Praise of Constantine*. In Drake, H. A. (trans.) *In Praise of Constantine: A Historical Study and New Translation of Eusebius' Tricennial Orations*, pp. 83–102. Berkeley, Los Angeles and London: University of California Press.
Jerome (1893). *Letters*. In Wace, Henry and Schaff, Philip (trans.) *The Principal Works of Saint Jerome: Nicene and Post-Nicene Fathers*, 2/6, pp. 1–295. Oxford and New York.
Paulinus of Nola (1966, 1967). *Letters*. In Walsh, P. G. (trans.) *Ancient Christian Writers*, 35, 35. New York and Ramsey, NJ: Newman Press.
Piacenza Pilgrim (1977). *Travels*. In Wilkinson, John (trans.) *Jerusalem Pilgrims Before the Crusades*, pp. 78–89. Jerusalem: Ariel and Warminster: Aris and Philips.
Sophronius of Jerusalem (1977). *Anacreontica* 19, 20. In Wilkinson, John (trans.) *Jerusalem Pilgrims Before the Crusades*, pp. 90–2. Jerusalem: Ariel and Warminster: Aris and Philips.
Theodosius (1977). *The Topography of the Holy Land*. In Wilkinson, John (trans.) *Jerusalem Pilgrims Before the Crusades*, pp. 63–71. Jerusalem: Ariel and Warminster: Aris and Phillips.
Valerius (1981). *The Letter in Praise of the Life of the Most Blessed Egeria Written to his Brethren Monks of the Vierzo by Valerius*. In Wilkinson, John (trans.) *Egeria's Travels to the Holy Land*, pp. 174–8. Jerusalem: Ariel and Warminster: Aris and Philips.

Secondary Sources (Citations in the text are according to page number.)

Avigad, Nahman (1983). *Discovering Jerusalem*. Jerusalem: Shikmona and Israel Exploration Society.
Avi-Yonah, Michael (1954). *The Madaba Mosaic Map*. Jerusalem: Israel Exploration Society.
Conant, Kenneth John (1956). The Original Buildings at the Church of the Holy Sepulchre in Jerusalem. *Speculum* 31/1: 1–48.
Danielou, Jean (1960). *The Bible and the Liturgy*. London: Darton, Longman and Todd.
Dix, Gregory (1986). *The Shape of the Liturgy*. London: A & C Black.

Eliade, Mircea (1959). *The Sacred and the Profane*, Trask, W. R. (trans.). New York and London: Harcourt Brace Janonovich.

Gadamer, Hans-Georg (1979). *Truth and Method*. London: Sheed and Ward.

Kutcher, Arthur (1973). *The New Jerusalem: Planning and Politics*. London: Thames and Hudson.

Limor, Ora (1988). The Origins of a Tradition: King David's Tomb on Mount Zion. *Traditio* 44: 453–62.

Nebenzahl, Kenneth (1986). *Maps of the Bible Lands: Images of Terra Sancta through Two Millennia*. London: Times Books.

Peters, F. E. (1985). *Jerusalem: The Holy City in the eyes of Chroniclers, Visitors, Pilgrims, and Prophets from the Days of Abraham to the Beginnings of Modern Times*. Princeton: Princeton University Press.

Van der Leeuw, G. (1986). *Religion in Essence and Manifestation*, Turner, J. E. (trans.). Princeton: Princeton University Press.

Vesely, Dalibor (1987). Architecture and the Poetics of Representation. *Daidalos* 25: 24–36.

Walker, P. W. L. (1990). *Holy City Holy Places? Christian Attitudes to Jerusalem and the Holy Land in the Fourth Century*. Oxford: Clarendon Press.

Wilkinson, John (1977). *Jerusalem Pilgrims Before the Crusades*. Jerusalem: Ariel and Warminster: Aris and Philips.

J. McKIM MALVILLE   JOHN M. FRITZ

Mapping the Sacred Geometry of Vijayanagara

INTRODUCTION

The Hindu empire of Vijayanagara was the largest and most effective in South Asian history, emerging at a time of considerable disarray in the south Indian political landscape. According to tradition the capital city of Vijayanagara was founded in 1336 CE by two warrior brothers of the Sangama family, Harihara I and Bukkaraya I. They constructed their capital city in an area which was already significant in local Hindu tradition but which had only limited political or economic significance.

Three successive dynastic families ruled the empire: Sangama from 1340 CE to 1486 CE, when control was wrested from the kings of the Saluva dynasty, who were themselves replaced by the Tuluva dynasty in 1503 CE, who then brought the empire to its apogee. The most effective rulers were Krishnadevaraya 1509–20 CE and Achutadevaraya 1520–29 CE. The city was sacked and burned following the disastrous battle of Talikota in 1565 CE.

At the time of its maximum power, Vijayanagara laid claim to approximately 360,000 km^2 and as many as 25 million people. Estimates of the population of the city itself range from 100,000 to 500,000, making it one of the major cities of the world at that time. The impregnable nucleus of the capital was protected by seven rings of massive walls and covered some 20 km^2. The surrounding urban environment encompassed more than 300 km^2.

The basis for this chapter is a series of measurements, undertaken during two field seasons, of the large scale organisation of the city and of orientations of temples, shrines and secular structures within it. On the basis of these measurements we find that Vijayanagara contains spatial organisation and symmetries not readily visible to the human eye. The Hindu temple appears to be a useful model for the city, partly because of the order and power contained in its centre, protected by extensive walls from the threatening chaos of the surrounding landscape.

The sacred hills of the city, empowered by creation mythologies, have greatly influenced its design. We are tempted to characterise the city's organisation as the result of a dynamic tension between traditional planning of Hindu cities as described in the architectural manuals, the *sastras*, and a dramatic but obdurate landscape which did not yield easily to the desires of city planners for centrality and symmetry (Fritz 1989).

Between its rounded granite hills, the major temples of the city appear to have been placed in unique linear relationships, similar perhaps in intent to the complex geometrical patterns of the Hindu temple which are invisible to normal human eyes but visible to the eyes of gods. The non-random placement of the major temples of Vijayanagara may have demonstrated some of the political power and sacred dimensions of the Vijayanagara kingship. The interlocking organisation of the city is itself a symbol of power, giving evidence of careful planning, as well as a homology between earth and cosmos.

The design of the city may also express the intent of the Vijayanagara kings to integrate or balance the various religious traditions of the empire. Within the royal centre of Vijayanagara we encounter a labyrinthine processional pathway, hidden by high walls from ordinary eyes, which in its microcosm may have provided a geometric integration of major sacred features of the city.

THE TEMPLE AND THE CITY

Vijayanagara was constructed as an imperial city at a site which had well-established mythological and epic traditions. The extensive urban area which emerged was a powerful vehicle for expressions of imperial power and covenants between gods and kings. Its size clearly demonstrated the authority of the Vijayanagara kings to mobilise massive labour forces. Its grandeur demonstrated an ability to attract south India's most talented architects and artisans.

Stein (1980) has suggested the Vijayanagara empire may best be understood as a 'segmentary state' consisting of relatively autonomous peripheral centres, dominated by a centre possessing great symbolic importance and ritual power. There appears to be little evidence for extensive administrative apparatus in medieval Hindu states or that the kings exercised extensive control over the

Mapping the Sacred Geometry of Vijayanagara

detailed affairs of the far-flung empire. The 'segments' of the empire were held together, Stein (1983: 75-8) proposes, in part by means of dramatic public rituals. Following the desacralisation of Hindu kings at the end of the Gupta age, divinity was transferred from kings to kingship, and public ritual became part of the basis for enhancing and confirming sacred kingship. Kings required continual divine sanction, and for that reason temples were lavishly supported by the kings, divinities from throughout the empire were incorporated in the city, and public festivals reaffirmed the link between kings and the gods.

Public ritual reached its most flamboyant expression in Vijayanagara at the time of the annual Mahanavami festival, 'great nine days', held in the bright half of the lunar month of ashvina (mid-September to mid-October). The events of the festival were centred at the 100 column hall and the great, truncated pyramid of the Mahanavami Dibba, at the summit of which the king viewed parades of priests, wives, elephants and soldiers (Stein 1983: 77-83). At times, the king shared his throne with a richly decorated, processional image of a deity. The king was at the centre of this demonstration of power, appearing as conquering warrior, possessor of immense wealth and celebrant wielding great ritual power. In addition to those numerous deities that had been imported from elsewhere in the empire and installed in the city, others were brought in during the festival and presented to the king. The city thus became a vast temple, surrounded by ramparts protecting king and images of deities. In fact, Stein (1983: 88-9) suggests that the most useful model for interpreting the Mahanavami festival is the South Indian temple, for at the time of that festival the palace and its precincts became a temple.

Based upon the spatial organisation of the city, Fritz (1985: 271-2) has also suggested that Vijayanagara may be understood as a temple, containing some of the spatial relationships of the classic Hindu temple. While the temple has many easily seen structures and symmetries, there is much hidden meaning and esoteric knowledge within its towers, corridors and sanctuaries. Everything is not put on public display, and indeed the essence of the temple is hidden in the dimly lit inner sanctum, the *garbagriha*, entered only by the Brahmin priests. Within the *garbagriha* is the symbol of the god imbued with divine power and knowledge. For the Hindu, knowledge is esoteric, buried to be recovered from the darkness of the cave, and the temple expresses that search for hidden knowledge (Shulman 1980: 19) adumbrated in its complex geometrical symmetries and mathematical patterns (Michell 1977: 73).

Spatial and Temporal Congruence

A guiding principle in our study of the sacred geometry of Vijayanagara has been that of *spatial and temporal congruence*. That principle affirms that patterns in space and time may acquire meaning and power through geometric similarity and synchronised parallelism to cosmic shapes and astronomical cycles.

The most frequently encountered example in India of congruence between heaven and earth, macrocosm and microcosm, is the Hindu temple with its frequent accurate alignment to the *true* four cardinal directions. Such congruency is additionally displayed by traditions that each temple is in the centre of the world, and by an architecture which mimics the cosmic mountain, Mt Meru, and the *axis mundi*. Within the city congruency may be evidenced by translational, rotational and reflective symmetries as meaning and power is transferred and bestowed from structure to structure. Furthermore, there is an economy of symbolism through the redundant use of certain forms and geometries. Only a few forms need to be 'memorised' and understood.

Through congruency, the attributes of kingship, priestly ritual, daily *pūjā* and public festivals were made parallel to the patterns of the macrocosm. As a consequence, human experience acquired characteristics of celestial archetypes and was made real.

The Mandala and the City

The paradigmatic example of spatial and temporal congruence is the *mandala*, a major regulating model for royal Hindu cities. In its purest geometric form, the mandala is a circle circumscribed by a square – the geometric pattern that represents the order of the universe. As the basic plan for a Hindu temple, the *vastupurushamandala* establishes the boundaries of the sacred enclosure and identifies both the point source out from which creation has emerged and the resulting ordering of the cosmos along cardinal directions (Kramrisch 1980). The centre of the *mandala* also represents the still point of the turning universe, the earth in a geocentric cosmos, and the surrounding square symbolises the path of the sun, moon and planets along the ecliptic.

The square of the *mandala* may be subdivided into a gridwork, symbolically interweaving all parts of the system into a whole. Each of the squares may be the seat of an important deity, with Brahma, the source of creation, often situated in the centre. The symbolism of an interconnected totality is an important attribute of the *mandala*. The universe thus defined consists of interlocking relationships and interdependent parts.

Sacred Hills

But the *mandala* is only part of the story of Vijayanagara's structure. The city is not a simple circle within a square in which one mountain, temple, or palace lies at its exact centre. Its powerful hills, invested with history, myth and sacred meaning, functioned as pivotal nodes in the organisation of the city. The city could not be a simple *mandala* independent of the local geography.

A. The highest hill, Matanga, dominates both the physical and symbolic landscape of the city and is woven by geometry and myth into nearly every one of its major features (Malville and Fritz 1993).

1. *Summit*: While the founding of Vijayanagara is historically credited to Harihara I and Bukkaraya I, the origin myths of the city identify the sage Vidyaranya as its founder. According to the *Rayavacakamu* (Wagoner n.d.), Vijayanagara was founded close to Matanga hill because of the symbolic protection promised by the hermit Matanga to Sugriva, the monkey king of the *Ramayana*. On the mountain, the natural predatory order was reversed: hares chased hounds; and all who dwelled upon it were protected from their enemies. Matanga hill was regarded as extending a symbolic blanket of protection over the surrounding landscape.

The ritual significance of the hill is demonstrated by three monumental staircases constructed of great slabs of granite, an apparent circumambulation (*pradakshina*) path around its base starting from the eastern end of the car street of Virupaksha temple, and the north road connecting the royal centre with the base of the mountain. Within the mountain are two sets of caves. Since caves and mountains are such fundamental aspects of Hindu mythology and temple design, their presence reinforces the symbolic and historic significance of the hill. The higher cave follows a semi-circular route within the mountain starting in an eastern facing, carved doorway and ending at a distance of 65 m in a small chamber containing a Shiva *linga* and a statue of Nandi. The larger and lower cave crosses the mountain from west to east starting with a carved image of Ganesh. Due to its size, the lower cave could have served as an excellent shelter in early times.

On the summit of Matanga hill stands the Virabhadra temple, dedicated to the fierce and militant form of Shiva who appears with four arms, holding a sword, shield, bow and arrow, and faces north. There are no inscriptions to date the construction of the temple, but the inner shrine of the temple is perhaps one of the earliest of the Vijayanagara period (Verghese 1989; Settar 1990: 24).

The temple lacks the architectural sophistication and symmetry of many of the Vijayanagara temples. Its somewhat haphazard and disjointed floorplan suggests several stages of construction and renders it difficult to establish definite axes. Our measurements of the columns in its eastern *mandapa* indicate an alignment to 359° 47'. Thus situated on the highest and most significant hill of the city, the shrine defers to no other point within the city and faces true north to within 13'.

Two major axes of the city cross on the summit of Matanga hill (Figure 1), a north–south axis which passes through Kodandarama temple on the north and through the large ceremonial gateway in enclosure IX of the Royal Centre, and a line connecting Matanga and Malyavanta hills which aligns approximately with sunrise on the morning of Makara *sankranti*.

2. *North–south axis* (NS): This axis divides the Royal Centre into two distinct sections that respectively manifest the public and private halves of kingship (Figure 2). Palaces are primarily in the western half, while the 100 column audience hall, the Mahanavami Dibba, the elephant stables and other structures associated with military activity lie to the east. The north–south line which crosses the tower of the Virabhadra temple on Matanga hill passes through the centre of one of the major gateways of the Royal Centre (Figure 2). As viewed from the centre of that gateway, the prominent *sikhara* of the Virabhadra temple lies less than 2' east of true north and, thus, directly above Matanga hill lies the fixed pole of the heavens. Since the unaided human eye can not resolve angles smaller than approximately 1.5', the accuracy of alignment is remarkable.

The motif of separation of space into eastern and western halves by the NS axis is further elaborated by the placement of the 100 column audience hall and the large palace in enclosure V (Figure 2). The NS axis nearly precisely bisects the line connecting them. The centre of the major, east facing room of the palace lies 92.8 m to the west of the axis while the centre of the 100 column hall lies 89.0 m to the east. As viewed from the mediating fulcrum midway between these two (public and private) structures, the *sikhara* of Virabhadra lies 4' from true north. Thus the very structure of Vijayanagara kingship appears to be sanctioned by the structure of the larger universe. The symbolism of north pole, cosmic mountain and *axis mundi* could not be more clearly presented. The sacred hill and its temple lie immediately under the pole of the

Mapping the Sacred Geometry of Vijayanagara

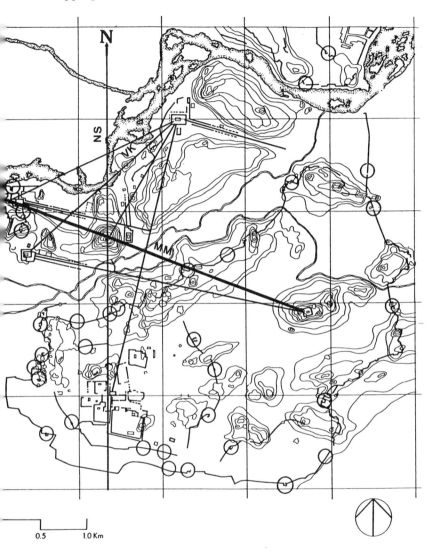

Figure 1 Axes of Vijayanagara: NS = north–south; MM = Matanga hill–Malyavanta; VK = Vithala temple–Krishna temple.

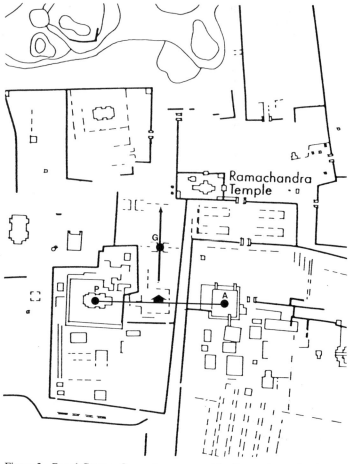

Figure 2 Royal Centre: G = gateway crossed by the north–south axis indicated by an arrow pointing to Matanga hill; A = 100 column audience hall; P = palace. The midpoint between palace and audience hall is marked by an arrow.

heavens as viewed from the ceremonial gateway in the royal complex.

3. *Matanga–Malyavanta axis* (MM): The most obvious manifestation of this axis is the location of Virupaksha temple (Figure 1). As measured from the roof of Virabhadra temple just above the shrine, the *goporum* of the Prasanna Malyavanta temple has an azimuth of 111° 35′. Exactly opposite lies Virupaksha temple; its

Mapping the Sacred Geometry of Vijayanagara 49

inner *goporum* has an azimuth of 291° 36', only 1' from 180° away from Malyavanta. The line connecting these three temples today approximately intersects with the rising sun on the morning of Makara *sankranti* when on January 14, the sun enters the constellation of Makaram. The first gleam of the rising sun occurs within a solar diameter of the *goporum* of Prasanna Malyavanta temple as seen from Matanga hill. The festival of Pongal occurring on this date has numerous dimensions and meanings, and one is a celebration of the turning of the sun back on a northern path. The difference between the current date of Pongal and winter solstice is due to the precession of the equinoxes. In the 16th century, Makara *sankranti* occurred approximately 7 days earlier and sunrise would have occurred within 2° of the line connecting the summits of Matanga and Malyavanta hills.

The MM line serves as an axis of symmetry (Figure 1) for Vithala and Ramachandra temples. Vithala lies 1.42 km to the north-east and Ramachandra lies 1.49 km to the south-west of the axis. These were the two pre-eminent temples in the city during the last 25 years of the empire.

4. *Rotation towards Matanga*: The organisation of the Royal Centre may have evolved as the kings shifted their allegiance towards Vaishnavite deities. The north–south axis dividing the Royal Centre may have originally reflected the mythic power of Matanga and his daughter, Pampa. Within the Royal Centre the Ramachandra temple, probably built by Devaraya I, 1406–22 CE (Verghese 1989), appears to have served as the state chapel during the Vaishnavite kingships. The temple has been rotated by 1° 30' away from true cardinality such that Matanga hill is framed in its northern gateway (Figure 2), and thus has sacred topography taken precedence over true cardinality.

A second and most remarkable example of rotation of a major temple away from cardinality occurs in Tiruvangalanatha, which has the curious property of possessing an east–west axis which is within 20' of true east–west, while its north–south axis is rotated to the east of north by 3° 10' (Figures 3 and 4). None of the other major temples of the city have such a large departure from perpendicularity of their major and minor axes. We suspect there is meaning hidden in such a skewed geometry. The orientation of the north–south axis is within 4' of the direction to the Hanuman shrine on the summit of Anjenadri hill *as seen from the summit of Matanga hill* (3° 6'). The major axis of Tiruvangalanatha temple appears to be the result of a nearly perfect geometric translation eastward from the summit of Matanga hill to its present location.

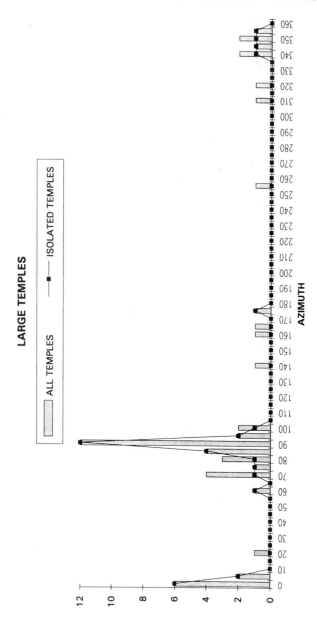

Figure 3 Distribution of orientations of temples. Those temples whose originations have not apparently been influenced by local geography or adjacent roads are identified as isolated.

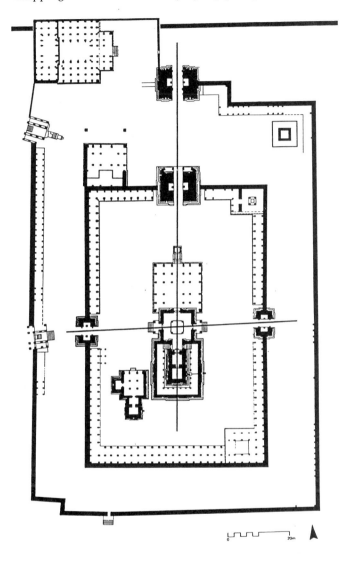

Figure 4 Tiruvangalanatha temple. Note the departure from perpendicularity of the major and minor axes.

Perhaps the symbolic protection of Matanga as well as mythological significance were transferred by such symmetry.

B. The prominent Malyavanta hill south-east of Matanga has two astronomical attributes in addition to its connection to the rising sun at the time of Makara *sankranti*. At summer solstice the sun rises over its summit as seen from the 100 column hall. Furthermore, crevices on its summit and the major temple, Prasanna Malyavanta, have orientations of 72–75°, which are close to the azimuth of the rising sun on the morning when it reaches the zenith.

1. *Summer solstice*: During the rainy season Rama retired to Malyavanta hill while Hanuman went southwards in search of Sita. As seen from the 100 column audience hall in the Royal centre, the sun rises above Malyavanta hill at the time of summer solstice. From the roof of the hall the king may have held audiences and observed the rising points of the sun above Malyavanta and along the east–west axis of the hall, occurring respectively at summer solstice and autumnal equinox. The rainy season starts soon after summer solstice and ends before autumnal equinox, and hence the arrival to and departure from Malyavanta by Rama parallels the movement of the sun as viewed from the 100 column hall.

2. *Zenith sun*: Before its association with Rama, the mountain was a sacred and important Shaivite site, named after a devotee of Shiva. On the summit to the west of the Prasanna Malyavanta temple there are 20 *lingas* and Nandis carved into the rock on either side of a crack which has an orientation of 74° 35′; the second crevice to the south of the temple has a similar orientation. In addition, the Prasanna Malyavanta and Lakshmi temples on the summit of Malyavanta hill have orientations of 72° 12′ and 72° 30′, respectively.

The mountain may have acquired additional symbolic relevance by its association with the sun on the day of the zenith sun, a highly important event in many societies which lie, like Vijayanagara, between the tropics of Cancer and Capricorn (Aveni 1981). The zenith sun is a visually dramatic phenomenon when vertical structures cast no shadows, which event occurs at the latitude of Vijayanagara near May 1 and July 26. On those days the azimuth of the rising sun is approximately 74°. There is no epigraphic evidence that the zenith sun was the occasion for an important festival in Vijayanagara; other than Makara *sankranti*, the majority of the festivals of Vijayanagara reported by Vergese (1989) were based on the lunar calendar. However, there is some evidence of attention to the zenith sun in Tanjore in Tamil Nadu, as local tradition indicates that there are days when the great temple of Brihadeshvara casts no shadow.

Our investigation of the orientations of over 100 small temples and shrines revealed further examples of potential alignments to the rising sun on the days of the zenith sun. Within that sample the only cluster of orientations which is statistically significant (lying 4.6 standard deviations above the mean) occurs between 70–80° (Figure 3). The large temples, by contrast, open primarily to the north and east.

Matanga at the Centre of the Mandala

Radiating outwards from Matanga hill are twelve lines which intersect and interact with significant features of the city (Figure 5; Table 1). Not only do they cast a web of interconnecting symbolism over the city, these radial lines also are expressive of the emergence and spread of order from a point source similar to that symbolised in the *vastupurushamandala*. According to the creation stories of Vijayangara in the *Rayavacakamu* (Wagoner n.d.), Matanga hill was indeed the source of the city. The major Hindu themes of creation and maintenance are thus manifested by this geometry with Matanga in the centre.

ORIENTATIONS OF TEMPLES

The construction of an individual Hindu temple involves the ritual of measurement in determining boundaries, orientation and proportion. Measurement established the demarcation of the ordered sacred space of the temple from unorganised profane space and was generally understood to be a re-enactment of the original measurement of the cosmos and its division into parts, thereby separating it from the dark waters of Chaos (Malville 1992: 26). Besides being well-proportioned, the temple had to be oriented to the true cardinal directions in order for it to achieve faithful parallelism with the cosmos and a successful departure from unmeasured, unbounded and unexplored space. The considerable attention, expressed as both theoretical justification and practical, quantitative technique, given to careful design and layout of the basic plan of the temple suggests that many major temples should be aligned to the true cardinal directions. The accuracy of alignment reflects not only the technical skill of the architects but also the commitment of the patrons to achieve architectural perfection.

The larger temples naturally fall into several classes according to the accuracy of their alignment to the cardinal directions. The distribution of the azimuths of the 42 large temples of the city is

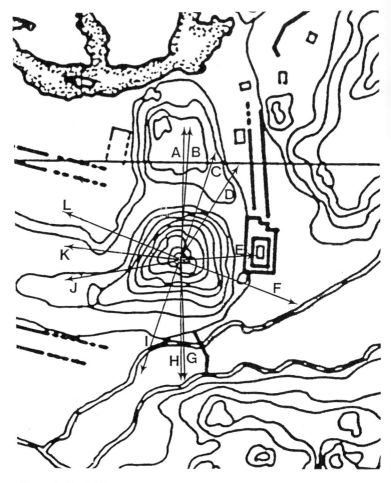

Figure 5 Radial lines from Matanga hill.

shown in Figure 3. Northern and eastern orientations are by far the most common, while an approximate westward orientation is represented by only one case. The ten large temples closest to true cardinality are listed in Table 2; the departures from cardinality of the first 6 temples average 14′.

INTERCONNECTIONS AND CITY PLANNING

As is indicated in Figure 1, the major temples of Vithala, Virupaksha, Krisha and Ramachandra appear to be contained in

Mapping the Sacred Geometry of Vijayanagara 55

TABLE 1: Radial lines from Matanga.

	Azimuth	Primary aspect	Secondary aspect
A	2° 2'	Chandrasekar	In line with Kodandarama
B	3° 6'	Anjenadri	In line with Kotilinga
C	16° 50'	Sugriva's Cave	Opposite Prasanna Virupaksha
D	35° 31'	Vithala	Symmetry with Virupaksha across VK axis
E	87° 10'	Tiruvengalanatha	Rotation to Anjenadri
F	111° 35'	Malyavanta	Opposite Virupaksha Sunrise Makara *sankranti*
G	178° 30'	Ramachandra	Rotation toward Matanga
H	180°	Enclosure IX	Ceremonial gateway 100 column hall/palace
I	196° 50'	Prasanna Virupaksha	Opposite Sugriva's cave
J	259° 30'	Krishna	Symmetry with Virupaksha across MH axis
K	274° 25'	Hemakuta	Rotation of gateway toward Matanga
L	291° 36'	Virupaksha	Opposite Malyavanta

TABLE 2: Temples closest to true cardinality.

		Central Temple		Departure from cardinality
		Axial length (m)	Area (m²)	
			A. 0–20'	
1.	Virabhadra	10	56	13'
2.	Chandrasekhar			14'
3.	Tirumangai Alvar	24	220	5'
4.	Vithala	71	1450	20'
			B. 30'–0°	
5.	Pattabhirama	70	1840	42'
6.	Krishna	47	660	47'
			C. 1–5°	
7.	Ramachandra	34	330	1° 30'
8.	Tiruvengalanatha	42	670	3° 10'
9.	Virupaksha	45	530	3° 52'
10.	Kodandarama	6	50	4° 43'

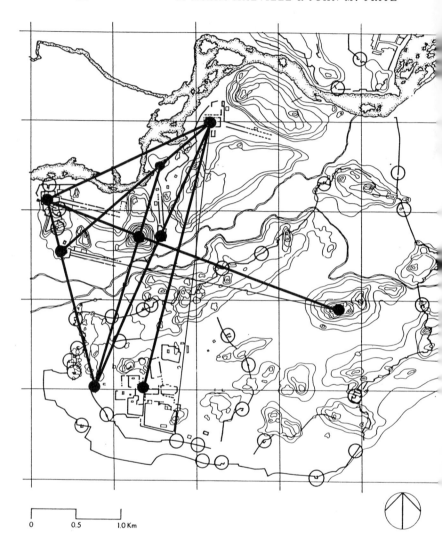

Figure 6a Organisation of the sacred sites of the city. The connection between the car street of Krishna temple with Malyavanta hill and that of the eastern section of the passageway in the Royal Centre with Vithala temple are also shown.

Mapping the Sacred Geometry of Vijayanagara 57

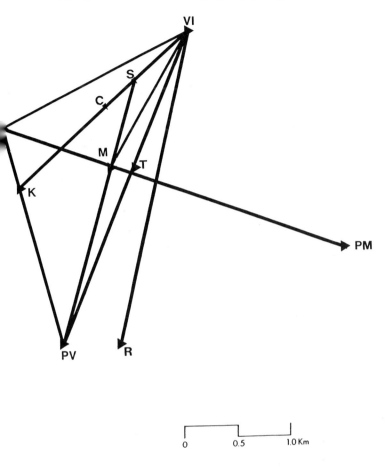

Figure 6b Identification of features: V = Virupaksha; K = Krishna temple; M = Virabhadra temple on the summit of Matanga hill; PV = Prasanna Virupaksha temple on the summit of Malyavanta hill; R = Ramachandra temple; VI = Vithala temple; T = Tiruvangalanatha temple; S = Sugriva's cave; C = Chakra *tirtha*.

an orderly pattern involving Matanga and Malyavanta hills. In Figure 6, where we include the remaining larger temples, a further order is revealed. Each of the larger temples has multiple connections with each other and/or features of the landscape.

Of all the major temples, Virupaksha was the first to be constructed and appears to be most completely determined by the natural landscape and lies at the intersection of a (somewhat vaguely defined) line between Hemakuta hill and the Tungabhadra river (Wagoner 1991) and the precise line containing Matanga and Malyavanta hills. An early form of the Virupaksha existed in pre-Vijayanagara times (Verghese 1989) as indicated by stone slabs in the Durga temple and the northern *gopurum* which contain dates of 1199 CE and 1230 CE respectively. Perhaps as early as the 7th century a temple was established on the point of intersections of these two lines. At that time orientation of its major axis may have been determined by the line connecting Hemakuta and Matanga hills.

The second major temple to have been built may have been Prasanna Virupaksha in the Royal Centre south of Matanga hill. The two natural features which may have determined its location are what is now known as Sugriva's cave and Matanga hill.

After Prasanna Virupaksha, Vithala may have been the next major temple to be established in the early part of the 15th century, perhaps as early as 1406 during the reign of Devaraya I, 1406–22 (Vergese 1989). Its location may have been established by an intended symmetrical relationship with Virupaksha and Prasanna Virupaksha. The distance between Vithala and Virupaksha is within 1 per cent of the distance between Virupaksha and Prasanna Virupaksha (VI–V: 1.98 km; V–PV: 2.02 km). There is also a reflective symmetry about the V–M–PM axis involving Vithala and Prasanna Virupaksha temples.

THE ROYAL LABYRINTH AND ESOTERIC CARTOGRAPHY

Within the Royal Centre there is a series of corridors that connect the Mahanavami Dibba with the Ramachandra Temple which are clearly distinguished by their non-cardinality and lack of right angles (Figure 2). Although we have no first-hand description of the use of this feature of the Royal Centre, this passageway may have been associated with a ritual procession from the Mahanavami Dibba to the Royal Chapel of Ramachandra temple. Proceeding in a clockwise direction, southward from the Mahanavami Dibba, the route paralleled the route followed by Rama (Fritz

1985: 268–9) as well as the direction of the episodes of the *Rāmāyana* depicted on the walls of the Ramachandra temple.

The first section which the king would have entered is tilted 4° east of north so as to be aligned with Vithala temple, some 3¼ km to north. The second section of the labyrinth is rotated nearly 11° south of east and is parallel to the car street of Krishna temple, and also the line connecting Krishna temple and Malyavanta hill. The actual numbers are impressive: the Krishna bazaar street has an azimuth of 101° 26' and the pathway has an azimuth of 101° 45'. Only the eyes of gods and the esoteric knowledge of priests and king could reveal the meaning of this elaborate hidden pathway.

And finally, the movement along the final stage would have carried the king towards both Ramachandra temple and the Hanuman shrine on the summit of Anjenadri hill, the birthplace of Hanuman. Even today, it is a dramatic moment when turning the south-west corner of the corridor, to view the brilliant white of the Hanuman shrine immediately above the *sikhara* of the Ramachandra temple. We cannot be certain that the Hanuman shrine was white at that time, but the other lines-of-sight involving the Hanuman shrine within the city suggest that it existed then and possessed a forceful symbolism. By moving within those carefully placed granite walls, the king could assimilate a good proportion of the sacred Vaishnavite geography of the city consisting of Vithala, Krishna and Ramachandra temples and Malyavanta and Anjenadri hills.

CONCLUSIONS

Because architecture controls the movement of people and limits vistas, we can believe that within the stones of the city are contained strategies for the enhancement and maintenance of power by the Vijayanagara kings. As we have noted, the city is not a collection of randomly sited structures scattered through an irregular landscape, but is a systematic whole with differentiated and interactive components. The model for such organisation of the city may be the Hindu temple with its concentration of power and order surrounded by protective walls. Such a city, containing both visible and invisible cosmological principles, succeeded for 250 years in infusing the Vijayangara kingship with divine power.

ACKNOWLEDGEMENTS

The field work described herein was performed while J.M.M. was a Fellow of the American Institute for Indian Studies. J.M.F. received funding from the Smithsonian Institution. We are grateful to the Government of India, the Archaeological Survey of India and the Karnataka Department of Archaeology and Museums for permission to work at Vijayanagara. At the archaeological camp at Kamalapuram, Sri Balasubramanya and his staff were of the greatest help. Drs George Michell, Philip Wagoner and Asim Krishna Das have provided valuable assistance and guidance on this project. Support in the field by graduate students Bernard Means and Frank Occhipinti is gratefully acknowledged.

REFERENCES

Aveni, A. (1981). Tropical Astronomy. *Science* 213: 161–71.

Dagens, B. (1984). *Architecture in the Ajitagama and the Rauravagama*. New Delhi: Sitaram Bhartia Institute of Scientific Research.

Fritz, J. M. (1985). Was Vijayanagara a 'Cosmic City'? In *Vijayanagara-City and Empire: New Currents of Research*, Dallapiccola, A. L. Vol. 1, pp. 257–73. Wiesbaden: Franz Steiner Verlag.

—— (1989). The Plan of Vijayanagara. Dallapiccola, A. L. (ed.) *Shastric Traditions in Indian Arts*, Vol. I, pp. 237–51. Stuttgart: Steiner Verlag.

Kramrisch, S. (1980). *The Hindu Temple*. Delhi and Varanasi: Motilal Banarsidass.

Malville, J. M. (1992). Cosmogonic Motifs in Hindu Temples. In Lyle, E (ed.) *Sacred Architecture in the Traditions of India, China, Judaism, and Islam*. Edinburgh: Edinburgh University Press.

Malville, J. M. and Fritz, J. M. (1993). Cosmos and Kings at Vijayanagara. In Ruggles, C. and Saunders, N. (eds) *Astronomies and Cultures*. Niwot: University Press of Colorado.

Michell, G. (1977). *The Hindu Temple*. Chicago: University of Chicago Press.

Settar, S. (1990). *Hampi, A Medieval Metropolis*. Bangalore: Kala Yatra.

Shulman, D. D. (1980). *Tamil Temple Myths*. Princeton: Princeton University Press.

Stein, B. (1980). *Peasant, State and Society in Medieval South India*. (Delhi).

—— (1983). Mahanavami: Medieval and Modern Kingly Ritual in South India. Smith, B. (ed.) *Essays on Gupta Culture*. Delhi: Motilal Banarsidass.

Verghese, A. (1989). *Religious Traditions in the City of Vijayanagara prior to 1565 AD*. Thesis, University of Bombay.

Wagoner, P. (1991). Architecture and Mythic Space at the Hemakuta Hill: A Preliminary Report. In Devaraj, D. V. and Patil, C. S. (eds) *Vijayanagara, Progress of Research 1984–87*. Mysore: Directorate of Archaeology and Museums.

—— (n.d.). *Tidings of the King: An Account of Krisnadevaraya of Vijayanagara*. Translated from the Telegu *Rayavacakamu* with an introduction. Unpublished manuscript.

EMILY LYLE

Internal–External Memory

According to a story told by Socrates in Plato's *Phaedrus*, when Theuth (the Egyptian god Thoth) introduces his invention of writing to Thamus (the god Amun), claiming that he has discovered 'an elixir of memory and wisdom', Thamus replies that, on the contrary, the invention will 'produce forgetfulness in the minds of those who learn to use it', and adds:

> Their trust in writing, produced by external characters which are no part of themselves, will discourage the use of their own memory within them. You have invented an elixir not of memory, but of reminding . . . .[1]

Similarly, Merlin Donald, in *Origins of the Modern Mind*, speaks of our bank of information arrived at through writing as External Symbolic Storage (ESS) and comments that nowadays the 'major locus of stored knowledge is *out there*' and that we carry around in biological memory a code rather than a mass of detailed information (1991: 314). If 'reminding' is a means of entering the total system, Donald can be said to agree with Thamus when he comments that many of the skills being taught today are 'memory-management skills' concerned with how to find, scan and assess (322). This is an end-result of the spread of the art of writing.

Donald postulates three stages in the evolution of culture and cognition between the time of the transition from ape to human and the present: 1) the mimetic, 2) the mythic, and 3) the theoretic, and he gives a strong sense throughout his book of developing symbolic capacity pushing back the frontiers to arrive first at language and then at writing. He suggests the possibility that the primary human adaptation was not language in itself but rather 'integrative, initially mythical, thought' which created the pressure to improve the conceptual apparatus (215), and that similarly the invention of writing was driven by conceptual needs (333). Donald argues that a new system of memory representation underlay each of the three key transitions in the course of human development and that, while the transitions to mimetic and mythic culture were dependent on new *biological* hardware in the nervous system, the third transition was 'dependent on an equivalent change in *technological* hardware, specifically, on external memory devices' (274). The capacity for memory first expanded within the brain and then externally through writing.

What developments in the direction of external symbolic storage were possible before writing? Donald speaks interestingly of an External Memory Field with a limited storage capacity which is 'a cognitive workspace external to biological memory' (296–7), and I think we can take it that the use of an external field which was compatible with the fullest use of internal human memory reached a peak in the absence of writing. When people were still dependent on the ability that was 'part of themselves', as Plato says, there was a symbiotic relationship between internal memory and external representations that can be called internal–external memory. With the introduction of writing, full use of this mode of thought became obsolete since writing was in many ways superior. However, writing itself is now being superseded by the symbolic potential of computers and so we have a double reason for grappling with this topic for, quite apart from the intrinsic interest of non-literate symbolic systems, we may find useful clues to future development through studying them (cf. Harris 1986).

Donald notes that the first pictorial images were already external representations that 'existed outside of the individual, rather than in visual memory' and that therefore 'a technological bridge was under construction that would eventually connect the biological individual with an external memory architecture' (284). He draws particular attention to the role of early devices that were analogue in nature, measuring 'one dimension of reality in terms of another' (337) and gives as an example of the practice of analogue visual modelling as used in societies without writing the case of the Walbiri in Australia who draw accompanying graphs in the sand when telling stories. For example, in a narrative about the travels of an ancestor, places may be shown by circles and connected with lines to indicate their paths and Donald comments 'this overall arrangement might be regarded as a prototypal map' (336). Nancy Munn, in the article quoted by Donald, points out (1973: 216) that circle and line carry a heavy load of symbolism and that visual perception of the circle–line configuration is only one aspect of a total process that articulates the relationship between the individual and the world order. The 'map' is not a simple one but is heavily charged with meaning.

Internal–External Memory

Shapes may be given external visual form in drawings and ceremonial objects which have the potential for remaining as permanencies (although they may also be quite ephemeral like the drawings referred to by Munn which are obliterated by dancing feet during a ritual celebration) or they may be expressed in dance or gesture which do not last beyond the moment. In any case, they do not exist independently of the inner concepts of such shapes which have as part of their content the journeys of imaginary ancestors moving from place to place. The places are all held together invisibly in the mind as a journey is imagined but the places themselves are real and visible. And here I think we need to develop what Donald has said considerably in order to understand the operation of what I have called internal–external memory in its cultural context. In particular, we have to keep in mind the point that humans make use of pre-existing shapes and objects on which no trace is left of human use. The situation has something in common with the discovery of an *objet trouvé* in art; if you can find something, there may be no need to make it. This is probably rather obvious in the case of a map when the route travelled is experienced or 'found'. If you have taken a journey you retain a mental representation of it in memory – a mental map which you may or may not choose to express by an external diagram which is an analogue of the journey. The same thing may happen in reverse, as it were; if you have concepts (like those of the Walbiri ancestors and their activities) that you wish to externalise you can do this by using real places without using a drawing.

The drawing, of course, has certain advantages. It is compact and may make the whole journey of the ancestor powerfully present in the small area of a ceremonial ground where people are assembled, and it is an aid to communication which is particularly useful as a teaching device in association with oral commentary when conveying knowledge to novices. Donald comments that, once visual analogue symbols like the Walbiri circle–line configuration had been created, 'the opportunity arose to use them as memory storage media' (338). I would like to look at this use, but would like to stress in addition that the real world unmediated by a drawing or chart can also serve for memory storage. The real world may be a *carte trouvée* where the human input is in the selection, not in the creation, of physical marks or objects.

One culture where the real world, diagrams and internal concepts are all in active use in a memory system is that of Puluwat in the Caroline Islands which is discussed by Per Hage in an article in *Oceania* (1978) and subsequently, in the context of graph theory, by Per Hage and Frank Harary in *Structural Models in Anthropology* (1983: 14–5, 19–20). The Caroline Islands are in Micronesia, which is well known for the feats of its navigators (cf. Hutchins 1983) whose traditional training enables them not only to make extended voyages in the open sea but also to move through conceptual worlds not exclusively concerned with the practicalities of navigation. We have the real world of the ocean and we have what the navigator carries around inside his brain. His chart, as he goes on voyages, is a mental one, and the mind is responding to physical places on the earth or sea or to stars in the sky without an intermediary. Nevertheless, the Micronesians employ humanly produced systems which can be laid out with pebbles on the ground, as is done during the teaching process (Gladwin 1970: 129–30). These systems are maps of the real world but incorporate also imaginary places. In one type of system, the trigger fish is used as model. This real fish gives an outline of a shape (Figure 1a) that is conceptualised in a particular way which is illustrated diagramatically by Hage (Figure 1b). Saul Riesenberg (1972: 32) explores the use of the model as explained to him by a Puluwat navigator as follows:

> In learning each Trigger Fish diagram, according to Tawuweru, you imagine yourself sailing west to the tail, under the setting of Altair, back to the centre (the backbone) and then due east to the head, then back to the centre and south under the Southern Cross to the ventral fin, back to the centre and north under the North Star to the dorsal fin, and finally back to the centre again.

In the system called 'trigger fishes tied together' (Figure 1c), eleven of the places are islands, reefs and a bank (fixed items such as would appear on standard maps), while Apilú is the personal name of a frigate bird seen in rough water, Kafeŕoor is a mythical vanishing island, Nalikáp and Nókitikiit are big waves, Máŕipeŕip (which means 'small pieces') is a very large, destructive whale and Fanuankuwel is the place of a whale with two tails.

Hage comments on systems like this (1978: 86–7) that they are not just maps for getting about but 'are mnemonic devices for the storage and retrieval of other kinds of cultural information – myths, spells, ceremonies, chants, recitations, etc.'. He adds:

> Each location in a sequence is a cue for an item of information and each sequence contains some homogeneous, ordered set of items. This type of device is not unique to the Puluwatese but is well known from the classical literature on rhetoric as the method of loci or the artificial memory . . . .

I shall return later to another Micronesian system, but shall look

Internal–External Memory

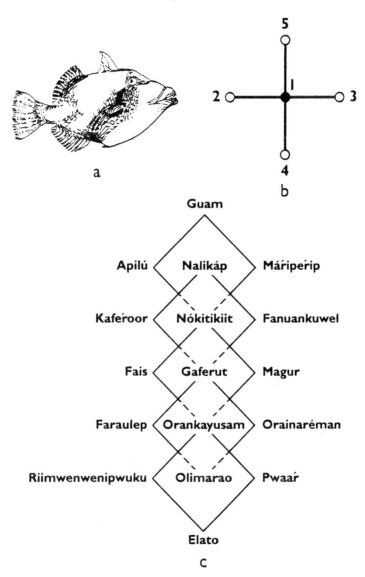

Figure 1 (a) Titan triggerfish (*halistoides viridescens*), after Randall et al. 1990: 453, cf. xix, 451–9; (b) after Hage 1978: 83; (c) after Riesenberg 1972: 31.

first at this method of *loci* or 'places' as discussed in three Latin works – the anonymous treatise on rhetoric addressed to Herennius dated *c.* 86–82 BC referred to as *Ad Herennium*, Cicero's *De oratore* dated 55 BC and Quintilian's *Institutio oratoria* of the first century AD. Although these treatises come from a literate society, they are concerned with a process that dispensed with writing for the orator delivered his speech without written notes and instruction in memory was intended to help him to recall the sequence of topics in his speech (the *res* or 'things'), with the option of also recalling words if he chose to incorporate this in his learning.

Quintilian gives the most detailed account of how the system was set up:

> Some place is chosen of the largest possible extent and characterised by the utmost possible variety, such as a spacious house divided into a number of rooms. Everything of note therein is carefully committed to the memory, in order that the thought may be enabled to run through all the details without let or hindrance. . . . The next step is to distinguish something which has been written down or merely thought of by some particular symbol which will serve to jog the memory; this symbol may have reference to the subject as a whole, it may, for example, be drawn from navigation, warfare, etc., or it may, on the other hand, be found in some particular word. . . . However, let us suppose that the symbol is drawn from navigation, as, for instance, an anchor; or from warfare, as, for example, some weapon. These symbols are then arranged as follows. The first thought is placed, as it were, in the forecourt; the second, let us say, in the living-room; the remainder are placed in due order all round the *impluvium* and entrusted not merely to bedrooms and parlours, but even to the care of statues and the like. This done, as soon as the memory of the facts requires to be revived, all these places are visited in turn and the various deposits are demanded from their custodians, as the sight of each recalls the respective details. Consequently, however large the number of these which it is required to remember, all are linked one to the other like dancers hand in hand, and there can be no mistake since they join what precedes to what follows, no trouble being required except the preliminary labour of committing the various points to memory.[2]

This account of Quintilian's is still referred to in memory studies today, for the point is that the system works. What we need is a fixed grid with a set order of succession laid down in advance and

Internal–External Memory

to this we can attach single items to be remembered which we can then pluck out of the system at will.[3]

I am concerned in this discussion particularly with the matter of what is internal and what is external and I think we have to realise that Quintilian is speaking of the actual practice of a Roman rhetorician and not of the total range of possibilities in the system. As Edward S. Casey stresses in a recent study of memory 'both the grid and its content are present as images' (1991: 127). When the orator gives his speech, he envisages within himself both the grid of places and the things he has put in the places. This point needs to be made very explicitly since the Latin words used are *loci* (places) for the grid and *imagines* (images) for the deposits. However, the Romans evidently recognised that both could have the status of images, as is made clear by Quintilian's added remark: 'What I have spoken of as being done in a house, can equally well be done in connexion with public buildings, a long journey, the ramparts of a city, or even pictures; *or we may even imagine such places to ourselves.*'[4]

Both places and deposits can be imagined, then, but what about their status in real-life existence? Quintilian says: 'We require, therefore, places, real or imaginary, and images or symbols, which we must, of course, invent for ourselves.' And here we have to allow for the fact that he is dealing with Roman practice, for the rhetorician setting up a memory system might choose a real building, like a house, but he confined his placing of deposits to the mental sphere and did not actually go about the house putting a physical anchor and weapon in the relevant positions. However, there is nothing to prevent the memory system operating through physical objects. Places and deposits can both be real (for example, a real anchor beside a real pillar) and they can both be images (for example, an imagined anchor beside an imagined pillar), or we can have real places and imagined deposits (for example, an imagined anchor beside a real pillar). There may be some greater difficulty in having real deposits and imagined places since real objects have to exist in space. In this case we would have to imagine away the real place and give a fresh imaginary context. For example, a real anchor could be seen on board a ship or placed on the ground during a ceremony and could be imagined as situated beside the imaginary pillar.

Although visual images seem to play a major part in this imagining, and I have dealt in terms of images up to now, it seems that there may also be a system within the mind which is not brought into visibility even introspectively. What is in the mind may or may not be visualised.

Casey first speaks of images in contrast to external reality but, as he develops his discussion, it becomes apparent that there can be imageless concepts also. He remarks that 'the spatiality which is operative in memory imagery cannot be reduced to strictly visual properties alone; it is not pictographic in nature, since it may include items not pictured at all' such as, in one example used experimentally, a harp hidden inside the torch of the Statue of Liberty, which could be recalled as well as one imaged as externally and visibly placed on the torch (1991: 129–30). Similarly, Alfred Binet, in a study of people playing blindfold chess (1894, 1966; cf. Paivio 1979: 24–6), found that there were varying degrees of use of mental imagery and that the more expert players made less use of imaging than novices. The expert user of a memory system may have employed comparable skills to those of a chess-master.

One memory expert from antiquity is known to have used a scheme of time rather than place, although, of course, the temporal can be mapped onto the spatial. He is mentioned in connection with the method of places by Cicero who says that he had himself met 'people with almost superhuman powers of memory' including Metrodorus of Scepsis, who was a rhetorician at the court of Mithradates Eupator, King of Pontus.[5] Quintilian expresses wonder that Metrodorus should have found three hundred and sixty places in the twelve signs through which the sun moves,[6] and L. A. Post has suggested (1931–2: 109) that he was familiar with the astrological practice of dividing the zodiac not only into twelve signs but also into thirty-six decans with associated figures, each covering ten degrees, and that he probably grouped ten places under each decan figure.

We have no evidence on whether or not Metrodorus split up his system into small sections in this way, but the practice of grouping is clearly attested in *Ad Herennium* where the author advises marking each fifth place and explains how this can be done: 'For example, if in the fifth we should set a golden hand, and in the tenth some acquaintance whose first name is Decimus, it will then be easy to station like marks in each successive fifth place.'[7] As the psychologist Allan Paivio comments (1979: 159, 212, 419), this passage illustrates the process of chunking or unitisation by which subjects remember material where the number of elementary units exceeds memory span by chunking them into higher-order units that can function as the memorial elements. He notes that G. A. Miller (1956) 'found that the immediate memory span was relatively constant at "seven plus or minus two" units over a wide range of materials and situations'.

I shall now return to the Caroline Islands and touch on the most esoteric system of all, which is called 'sea-life'. In this system, each inhabited island has imaginary lines running outwards from it towards twenty-eight of the thirty-two navigational star positions used by the navigators, the four omitted ones (the rising and setting of alpha and beta Aquilae) being very close to the rising and setting of Altair and being considered the wings of the big bird of which Altair is the body. It seems that the system was thought of in quarters since the quadrants are indicated in the circle of pebbles representing navigational star positions and since Gladwin states that a square was formerly in use before the introduction of the circle (1970: 129–30), and this means that we have a case of a memory system that demonstrates chunking by sevens. The seven places in the south-east quadrant about which Gladwin gives some information are the rising or eastern positions of Altair, Orion's Belt, Corvus, Antares, Shaula, the Southern Cross at rising, and the Southern Cross at 45°. Running in these directions from Puluwat are the sea-life lines along each of which is located every few miles an item which is generally imaginary but is occasionally a real place like a reef or an island. Although the twenty-eight places of the complete series are derived from the sky, they are locatable also in familiar settings through the identification of landmarks as described by Gladwin (180): 'Every navigator knows where each of the navigation stars sets on Allei as seen from his own canoe house: Orion's Belt by the old dead tree; Altair over the half-sunken ship; Aldebaran between the two tall breadfruit trees; and so on.' We can see clearly in this system the three levels of: star positions or landmarks (human selection), placed pebbles (human artefact) and mental locations (human interior conceptualisation).

I suggest that any complete study of human development must take account of the very considerable capacity for memory storage in the absence of writing through the interplay of visible and invisible worlds. Frances Harwood has argued (1976: 795) 'that the precise geographical locations given with such frequency in the myths of nonliterate societies are not mere embellishments, but play a significant role as mnemonic devices for the recall of the mythical corpus', and, similarly, points in the year cycle may serve as temporal 'places' for the deposit of information to be remembered, as the case of Metrodorus has illustrated. Structurings of place and time should be seen, along with the structuring of speech, as factors in the establishment and retention of a cultural reservoir in what Donald calls 'oral–mythic' society. This cosmological component needs to be brought to the fore and integrated into current discussion.

NOTES

1 *Phaedrus* 274–5, trans. Fowler 1914: 562–3.
2 This and the following quotation are from XI.ii.21, trans. Butler 1921–2: 4.222–3.
3 Although the grid is laid down in a set order of succession, it should be noted that, as Paivio comments (1979: 159): 'this is no sequentially constrained system – the memory information is spatially parallel' with the result that we can 'name the images in our ordered memory loci in any order'.
4 Butler begins a new sentence at the second 'or'. The italics are mine.
5 II.lxxxviii.360, trans. and ed. Sutton and Rackham 1942: 1.470–1. For discussion of Metrodorus, see Yates 1966: 19–20, 23–5, 39–44.
6 XI.ii.22, trans. Butler 1921–2: 4.224–5.
7 III.xviii.31, trans. Caplan 1954: 210–11. I have changed Caplan's word 'background', translating '*locus*', to 'place'.

REFERENCES

Binet, Alfred (1894). La mémoire des joueurs d'échecs. Part 2 of *Psychologie des grands calculateurs et joueurs d'échecs*. Paris: Hachette.
—— trans. Marianne L. Simmel and Susan B. Barron (1966). Mnemonic Virtuosity: A Study of Chess Players. *Genetic Psychology Monographs* 74: 127–62.
Casey, Edward S. (1991). *Spirit and Soul: Essays in Philosophical Psychology*. Dallas, Texas: Spring Publications.
Cicero, M. T., trans. E. W. Sutton and H. Rackham (1942). *De Oratore*. 2 vols. Cambridge, Mass.: Harvard University Press.
Donald, Merlin (1991). *Origins of the Modern Mind: Three Stages in the Evolution of Culture and Cognition*. Cambridge, Mass.: Harvard University Press.
Gladwin, Thomas (1970). *East Is a Big Bird: Navigation and Logic on Puluwat Atoll*. Cambridge, Mass.: Harvard University Press.
Hage, Per (1978). Speculations on Puluwatese Mnemonic Structure. *Oceania* 49: 81–95.
Hage, Per and Frank Harary (1983). *Structural Models in Anthropology*. Cambridge: Cambridge University Press.
Harris, Roy (1986). *The Origin of Writing*. London: Duckworth.
Harwood, Frances (1976). Myth, Memory, and the Oral Tradition: Cicero in the Trobriands. *Oceania* 78: 783–96.
Hutchins, Edwin (1983). Understanding Micronesian Navigation. In Gentner, Dedre and Stevens, Albert L. (eds)

Mental Models, pp. 191–225. Hillsdale, NJ and London: Lawrence Erlbaum Associates.

Miller, George A. (1956). The magical number seven, plus or minus two: some limits on our capacity for processing information. *Psychological Review* 60: 81–97.

Munn, Nancy D. (1973). The Spatial Presentation of Cosmic Order in Walbiri Iconography. In Forge, Anthony (ed.) *Primitive Art and Society*, pp. 193–220. London and New York: Oxford University Press.

Paivio, Allan (1979). *Imagery and Verbal Processes*. Hillsdale, NJ: Lawrence Erlbaum Associates.

Plato, trans. H. N. Fowler (1914). *Euthyphro, Apology, Crito, Phaedo, Phaedrus*. London and New York: William Heinemann and Macmillan.

Post, L. A. (1931–2). Ancient Memory Systems. *Classical Weekly* 25: 105–10.

Quintilianus, M. F., trans. H. E. Butler (1921–2). *The Institutio Oratoria of Quintilian*. 4 vols. London and New York: William Heinemann and G. P. Putnam's Sons.

Randall, John E., Allen, Gerald R. and Steene, Roger C. (1990). *Fishes of the Great Barrier Reef and Coral Sea*. Honolulu: University of Hawaii Press.

Rhetorica ad Herennium, trans. Harry Caplan (1954). London and Cambridge, Mass.: William Heinemann and Harvard University Press.

Riesenberg, Saul H. (1972). The organisation of navigational knowledge on Puluwat. *Journal of the Polynesian Society* 81: 19–56.

Yates, Frances A. (1966). *The Art of Memory*. London: Routledge and Kegan Paul.

MARK NUTTALL

Place, Identity and Landscape in North-west Greenland

Since the introduction of Home Rule in 1979, the political, cultural and economic development of Greenland has been described as a process of 'nation-building' (Dahl 1988), underpinned ideologically by policies of 'Greenlandisation' (Grønlandisering) which aim to develop the country in terms of its own conditions and available resources. A sense of national Greenlandic identity has been fostered and political rhetoric emphasises a homogeneous Greenlandic culture (Nuttall 1992). Ethnopolitical symbolism has done much to strengthen the institutionalisation of a national Greenlandic identity (Kleivan 1991), but the emerging diversity of occupational associations and the beginnings of structural differentiation complicates the discussion of contemporary Greenlandic identity and raises questions about how it can be studied and represented. In this respect, anthropological research on the construction of Greenlandic identity has, as Petersen (1991: 21) puts it:

> ... been critical in the sense that it has commented on the potential intolerant tendencies inherent in a too narrow focus on the content of identity or the content of 'Greenlandicness'. It may be argued that anthropologists have helped turn the perspective from heritage to culture.

Peterson argues that the identity debate has both a cultural and a structural dimension. The cultural dimension, espoused in particular by the ruling Siumut party,[1] is characterised by specific cultural and ethnic aspects such as language, occupation, genealogy and birthplace as defining components of Greenlandic identity. Identity as a Greenlander tends, therefore, to be regarded as an Inuit identity vis a vis Danes. A Greenlandic identity with a structural dimension, on the other hand, is broader, encompassing both ethnic Greenlanders (i.e. Inuit) and Danes living in the country. It also opposes Greenlanders to other Inuit (Petersen 1991).

This chapter does not intend to add to the debate on structural and cultural dimensions of identity, or concentrate on the content of 'Greenlandicness', but it does take the position that the definition and working out of identity in Greenland is contextual. People construct their cultural worlds and their identities from significant centres that have meaning for them, such as kinship, naming relationships, mode of production, the environment, locality and place, and the way people articulate their cultural identities finds expression in many ways, including various levels of metaphor, myth and narrative. Much of this has been discussed in more detail elsewhere[2] and this chapter focuses on aspects of the significance of place for cultural identity in the seal hunting settlements in the Upernavik district of north-west Greenland, by concentrating on cultural understandings of the landscape as expressed in place names and the meanings behind them.

In this way, the discussion of cultural identity is approached from the perspective of locality and the significance and specificity of place, illustrating how the landscape is used and understood in a local sense rather than understanding landscape from a national, political and wider territorial perspective. In the settlements of the Upernavik district, people regard the environment not only as providing their immediate needs,[3] but as an expression of and metaphor for community. Attachment to the environment is both sentimental and symbolic. There is a powerful feeling of belonging; naming and memory show how places have a far deeper meaning than is immediately apparent.

THE CULTURAL USE OF SPACE IN NORTH-WEST GREENLAND

The Upernavik district of north-west Greenland extends from the Svartenhuk Peninsula at 71° 28′N up to Melville Bay in 75° N. The district's Greenlandic Inuit population of 2149 lives in the administrative centre of Upernavik town and ten outlying settlements. While a small-scale commercial fishing industry has developed in the district in recent years (Forchhammer 1989) the hunting of sea mammals, mainly seals, continues to underpin the subsistence economy of the settlements (Nuttall 1992).

Much of the literature on Inuit subsistence systems, in Greenland and in other parts of the Arctic, has focused on the cultural use of space, as opposed to the cultural meaning of place. Haller's study of Nuussuaq, in the northern part of the Upernavik district, concentrated on that settlement's annual subsistence resource area, or '. . . the total resource area utilized by hunters during the annual cycle of hunting activities' (1986: 43), what Haller also refers to as 'action space'. Focusing on spatial aspects of the marine hunting culture throughout Upernavik district, Haller examined the delimitation of boundaries, the overall settlement pattern, characteristics of settlement location, mapped activity patterns, and looked at pattern change as an indication of culture change.

In this way, Haller was concerned with how subsistence activity patterns contribute to the establishing and defining of cultural boundaries. Furthermore, through the study of activity patterns he suggested that a culture system and cultural groups within that system could be identified. Activity patterns have also been considered important for understanding the cultural use of space by Inuit elsewhere in the Arctic (e.g. see Freeman 1976), especially as land claims have been at the very centre of Inuit self-determination movements.

Haller's study, while concluding that hunting activity and spatial organisation in Upernavik district has remained consistent over time, was nonetheless objectivist and etic with regard to its emphasis on hunting as a biobehaviour system and the annual resource area as 'action space' (1986: 43). The ideal of empiricist inquiry into Arctic hunting cultures continues to be held by those who espouse evolutionary ecology and defend optimal foraging theory (e.g. see Smith 1991), yet while unaffected by post-modernist leanings towards more literary ethnographic writings, such studies do tend to leave out the person of the hunter and the distinctive ideas and cultural meanings that people attribute to their environment. By describing the environment of the hunter as action space, Haller ignored the cultural and symbolic dimensions of landscape that make 'thought space' or memoryscape (Nuttall 1991) more accurate terms to use when discussing how local areas are perceived and thought about by those living there.

THE CULTURAL MEANING OF PLACE

The objectivity of the theoretician in studying the cultural use of space, however, has indeed characterised much of the writing by social scientists, including geographers as well as anthropologists, who work on concepts of place and region (Entrikin 1991). Entrikin, basing his discussion on the work of Nagel (1986), has argued that this decentred approach to the study of place widens the '... divide between the existential and naturalistic conceptions of place' (Entrikin 1991: 134). However, even a centred approach can veer dangerously close to a particularism that, while capturing the significance of place for individual attachment, nonetheless fails to retain an objectivity necessary for the discovery of the universal qualities of place. To understand the meaning of place, therefore, it is necessary to adopt a view from '. . . a point in between, a point that leads us into the vast realm of narrative forms' (Entrikin 1991). In this way, suggests Entrikin, it is possible to understand what it means to be both 'in a place' (centred and particularist) and 'at a location' (decentred and universalist).

In approaching place as location, scientific description and generalisation can obscure the cultural meanings individuals attribute to place. The relationship an individual has to a specific place is vital for both the cultural construction of self and for the construction of community and collective identity. Indeed, place is an important referent of a society's identity, constructed from experience and a continuous cultural and moral exchange between people and landscape (Tuan 1975; Weeden 1992). There is nothing inherent in a place, however, that makes it any more or less significant than another. A place is special because of:

. . . the way it buries itself inside the heart, not whether it's flat or rugged, rich or austere, wet or arid, gentle or harsh, warm or cold, wild or tame. (Nelson 1989: xii)

With this in mind, the approach favoured in this chapter is with place as context, an approach that argues for attempts to grasp the cultural richness of a specific place and the significant meanings places have in an existential and more immediate rather than universal sense. People invest places with individual and collective meaning and an anthropological understanding of this does not necessarily have to lean towards narrow particularism. By focusing on experiential aspects of culture, understanding the significance of the meanings individuals attribute to places in their locality is no different from '. . . grasping a proverb, catching an allusion, seeing a joke' (Geertz 1983: 70). But in order to grasp these meanings it is often a matter of recognising what is hidden in the landscape, beyond what is immediately expressed and perceived.

COMMUNITY AND LOCALITY

Throughout Upernavik district, individual membership of a particular community and an attachment to a specific locality are important referents for cultural identity. Community and locality are significant centres around which individuals focus their lives, participate in, and derive meaning from. In response to the profound economic and social changes that occurred throughout the Arctic during the 1960s and 1970s, young, emerging Inuit elites looked beyond a purely localised identity and sense of place towards wider national identities that championed the cause of

Inuit as a people rather than as diverse groups in space and time. Dahl (1988) regards the present Inuit ethnic groups in Greenland, Alaska and Canada as imagined communities of recent ideological construction. This is in contrast to traditional Inuit societies which did not have distinct names, but which referred to individuals and groups by geographical location rather than other characteristics such as social organisation. The identification of individuals and groups with specific localities persists in contemporary times, despite the use of wider ethnic terms such as Kalaallit, Inuit and Inuvialuit, which have the potential to obscure the diversity within the Inuit area.

In Greenland, for example, a person is still identified in relation to their home town or settlement. Identity and sense of place is most immediately expressed by using the suffix -*miut*, meaning 'people of' (singular -*mioq*). For example, in Upernavik district, people who live in the settlement of Kullorsuaq identify themselves as Kullorsuarmiut, i.e. 'people of Kullorsuaq' (sing. Kullorsuarmioq). In this way, they express their attachment to the locality and distinguish themselves from the people of Upernavik town (Upernavimmiut), Tasiusaq (Tasiusarmiut), Nuussuaq (Nuussuarmiut), and so on.

Hunting settlements in Greenland enjoy a localisation of resources and, because fixed settlements have been characteristic of Greenlandic Inuit society, groups of people have remained in the same areas rather than travelling over large parts of the country according to seasonal variation or the migration of animals. The continued use of particular hunting areas and campsites means each community has its own recognised territory, commonly known as *piniarfik* (the hunting place).

In Upernavik district subsistence activities take place in a specific environment; a pattern of islands, peninsulas and fjords, the latter being either ice-free or filled with icebergs calved from tidewater glaciers. Apart from some caribou hunting in the southern part of the district during the nineteenth and twentieth centuries, and a little fox trapping and ptarmigan hunting in present times, the population of Upernavik district has been concerned primarily with sea-use. Haller's point that the land 'cannot be considered a resource base' (1986: 97), however, should not mislead us into assuming that the landscape has no significance in a different sense.

The Upernavik environment is ordered and conceptualised by Inuit through the interaction of imagination, thought, experience and language.[4] The way people think and talk about their local landscapes reveals a complex knowledge not only about the physical environment but of real, mythical and imagined past events that have significance for the cultural construction of community. One way of exploring the cultural and symbolic dimensions of the landscape is to study place names and the meanings and stories attributed to them. Place names are multi-layered; they inform us about past and present subsistence activity and about a multiplicity of close associations between the human and natural worlds (Kleivan 1986; Nuttall 1992).

NAMING AND THINKING ABOUT PLACES

Throughout the Arctic, early European and American explorers named vast areas of territory in order to establish either a claim to possession by the nation states that sent out their expeditions, or in order to honour sponsors, kings, queens and even themselves. But in doing so, they ignored the existence of indigenous Inuit names that bind the landscape with human imagination and experience. British explorers, such as William Baffin in the seventeenth century and Sir John Ross in the early nineteenth century, named parts of the West Greenland coast during their expeditions to seek out and discover a North-west Passage, a corridor to the wealth, fortunes and dreams of Cathay. Later in the nineteenth century, American explorers such as Elisha Kent Kane, Isaac Hayes and Charles Francis Hall ventured into north-west Greenland in attempts to reach the North Pole and place names such as Cape Constitution, Washington Land, Harvard Islands, Kane Basin, Hall Land and Hall Basin are a legacy of their journeys.

To name places in this way is to reduce the landscape to uncharted wilderness, thought to be devoid of life. For explorers and colonists in the Arctic the land was simply awaiting 'discovery' and to be claimed for the countries they represented. In the Upernavik district examples of places named by Danish and English explorers include Holm Island, J. A. D. Jensen Islands, J. P. Koch's Land, and Sanderson's Hope. These are places named after explorers, prominent figures and expedition sponsors. For example, Sanderson's Hope was named by the Elizabethan seaman John Davis for William Sanderson, one of the sponsors of Davis' 1587 expedition to discover the North-west Passage. During the Danish colonial period in Greenland[5] the names given to places by English, American and Danish explorers were systematised by an official policy of using Danish names and terms for the geography of the entire country (Vaughan 1991: 174). In this way, for example, the Danish 'bugt', 'ø', 'sund' replaced the English equivalents 'bay', 'island' and 'sound'.

Places in Greenland were not only named by explorers and

Danish colonial agencies, however. The waters of west and north-west Greenland were regularly exploited by European whalers. With its origins in Basque whaling, the European whale industry had developed into the competitive 'Greenland fishery' for Right whales between the English and the Dutch off the coast of Spitzbergen in the seventeenth century. This resulted in over-exploitation of the bays of Spitzbergen and the Greenland Sea and led, in turn, to the seeking out and discovery of new waters in Davis Strait and Baffin Bay between Greenland and the Canadian East Arctic, where Right whales thrived in abundance. In pursuit of the Right whale, Scottish, English and Dutch whalers became regular visitors to the coasts of Baffin Island and Greenland, making contact and trading with the Inuit at their settlements and hunting camps. Evidence of the presence of these whalers is still to be found in the English and Dutch names given to harbours and other places along the west coast of Greenland (e.g. see Gulløv 1987).

Throughout the eighteenth and nineteenth centuries, Scottish and English whalers systematically exploited the waters in Upernavik district and are responsible for several place names, such as Sugar Loaf, Dark Head and Horse Head. As well as leaving this cultural imprint on the Upernavik landscape, the whalers often left behind their dead in lonely graves (e.g. see Nuttall 1990a). The whalers, as indeed the explorers, took their own cognitive stereotypes of the Arctic environment with them on their voyages. Idiosyncratic impressions were allowed to colour published journals of whaling voyages. Writing was influenced by experiences and by the author's situation, such as the hardships endured, cold, inadequate equipment, poor diet and health, homesickness, starvation and the death of fellow crew members. The environment of north-west Greenland was regarded as an unforgiving wilderness and Melville Bay in the northern part of Upernavik district became known as the whaleship's graveyard, with specific place names, such as Devil's Thumb[6] in the southern part of Melville Bay, reinforcing this image.

The names given to places in the Upernavik district by the indigenous Inuit, however, differ from the Danish and English names given to the same localities. This is also true for the rest of Greenland (Petersen 1985). Rather than being merely geographically descriptive, Greenlandic place names are multidimensional in that they contain physical, historical, factual and mythical meanings. In north-west Greenland a community's local area is a memoryscape of events and experiences (Nuttall 1991), with place names having '. . . a semantic depth that extends beyond the concern with simple reference to location or to a single image' (Entrikin 1991).

In Greenland it is rare to find places that Inuit have named after people, unlike the places named by the Danes such as Knud Rasmussen Land, Lauge Koch Kyst, Julianehab, Jakobshavn and Kong Frederik IX's Land. When Inuit do name places after people, the names usually refer to specific family camp-sites where individuals have community allocated rights and exclusive tenure, providing they continue to use the sites. If use is discontinued, then the rights revert to the community and another person is then free to take the site over. In this way 'Johannsi's camp-site' can become 'David's camp-site', illustrating how names in a landscape are not necessarily permanent.

Many Greenlandic place names in the Upernavik district are geographically descriptive, and may in fact correspond to the Danish names later imposed on the landscape. Examples include Saattut ('the flat islands', or Fladøerne in Danish), Avannarleq ('north island', or Nordø in Danish) and Sioraq ('sand island', or Sandøen in Danish). Other Greenlandic names which refer specifically to physical features include Kangeq ('headland'), Qerqertaq ('island'), Kullorsuaq ('big thumb'), Kangerlujssuaq ('big fjord'), Uummannaq ('mountain shaped like a heart'), Ikerasaq ('sound' or 'channel'), Itilleq ('the crossing place') and Majuariaq ('hill').

Certain place names reflect analogy and remind people of other realms of experience and culture which transcend geography. Examples include Amarortalik ('the place of wolves'), Maniitsoq ('island resembling hummocky ice'), Iviangernat ('twin peaks resembling a woman's breasts'), Toornaarsutoq ('little spirit' or 'the place with lots of spirits'), Angakkussarfik ('the place of initiation for a shaman'), and Tunisivik ('an offering place').

The majority of place names in north-west Greenland, however, inform us of subsistence activities. By recording place names, a comprehensive understanding of land and sea use, both in the past and present, can be gained. In this way a picture emerges of a landscape with features named not for any geographically descriptive reason, but owing to the importance of these places for hunting, fishing and other subsistence activities. Examples include Eqaluit ('char' or 'place of the char'), Arfeq ('whale', or 'the place where whales are caught'), Miteq ('eider duck', or 'the place where eider ducks are found'), Qasigiaqarfiup Kuua ('river which belongs to the place of the spotted seal'), Ukalersalik ('a place where one finds Arctic hare'), Qajartoriaq ('a lake where kayaks are used') and Tupersuusat ('a camping place').

But while these place names tell us about geographical description and subsistence activity, many have additional historical, symbolic and mythical meanings. That which is hidden and invisible in the landscape is revealed through stories and myths which unfold against a geographical backdrop. Stories may take the form of real and remembered things, and relate to specific events and experiences such as 'Maniitsoq is the place where I once caught a polar bear', or 'Qerqertaq is the place where my father used to spend the summer in a fishing camp'. The point to be made is that place names do not necessarily reflect individual and communal experience. Some place names may be mnemonic devices that trigger a collective memory of an event that has significance for the community as a whole, such as 'Kinnguvik ("the place where one capsizes") is the place where hunters used to capsize in their kayaks', and then stories would be told of hunters losing their lives there. Or particular places, such as Maniitsoq (near the settlement of Kangersuatsiaq) remind people of the time a manipulative, evil shaman was said to have lived in the area.

Alternatively, places have a more individual meaning and remind people of personal experience, such as 'Umiiarfik "a place for harbouring umiaks" is a dark, frightening and evil place because I had a dream about people living under the earth when once camped there'. It is impossible to record the numerous place names and the stories about places in the Upernavik district in a short chapter, but place names are important in storytelling in the settlements of north-west Greenland because they situate events in the physical setting where those events happened (Basso 1984: 32). Storytelling related to places means those places become remembered and provide a direct link between the past and the present. A myth, a story, a metaphor, or an allegory related to a place '. . . distils essences from generations of experience and adaptation' (Weeden 1992: 146). Just as spirits and other supernatural creatures that 'people' the landscape are metaphors that make the natural world intelligible, so place names, stories and events located in the landscape express and give meaning to the physical environment as a central feature of the cultural world of the Inuit.

Because membership of a community is important for identity, places are given meaning by those who live in a particular community. In Upernavik district hunting activity is 'carried out within de facto environmental boundaries' (Haller 1986: 146) and intimate knowledge of places does not usually extend to other community hunting areas. Changes in place names that are suffixed with -*kassak* ('bad' i.e. 'of no significance to hunting') usually denotes a cultural and symbolic boundary at which subsistence activity stops. An area that may be familiar to those who live, hunt and travel there becomes unknown territory to those who have no knowledge, experience or memory of it.

This last point illustrates the local nature of landscape in northwest Greenland. People perceive the physical landscape in a particular way and are involved in a dialogue with a local landscape that is suffused with memory and highly charged with human energy. Through the use of sea and land, and through myth and historical events an image of the community is reflected in the landscape. Places then resonate with community consciousness – they become reference points. Such a local landscape finds expression in the cognitive maps used by individuals to orientate themselves and their sense of community to it. The significance of a place for personal, family or community life is often difficult for outsiders to grasp, however, partly because of a need to explain, justify, translate and compartmentalise the experience of others into another system of logic. For example, an elderly hunter from the settlement of Kangersuatsiaq told me how, having been on a hunting trip, he had discovered his footprints at an old campsite of his that he had last visited twenty years previously. To misunderstand the significance of this is to misunderstand that there are many layers of meaning in what, to this man and others like him, is a cherished and known landscape. To understand this is to recognise that '. . . one world contains glimpses of others' (Okri 1991: 10).

The landscape is constituted in relation to each individual and through experience, knowledge and memory, images and understandings are constructed and negotiated (see Basso 1984). Stories are often personal, they chart the adventures and experiences of individuals. A story told by someone about finding his footprints in the land, or about the time he caught two polar bears on one hunting trip, for example, is something that Lopez regards as bringing together an external landscape and an interior one. The external landscape is the one we most immediately perceive through the senses, through sight, sound, touch and smell. But it is a landscape we get to know through understanding and perceiving the complexity of relationships in it. The interior landscape is within the self, a landscape of the mind, '. . . a kind of projection within a person of a part of the exterior landscape' (Lopez 1989: 65). Intuitions, ideas, experiences and memories belong to a set of relationships of the internal landscape that are sometimes very obvious and at other times very subtle and elusive to grasp. Lopez regards the internal landscape as responding '. . . to the character and subtlety of an exterior landscape; the shape of the individual

Place, Identity and Landscape in NW Greenland 85

mind is affected by land as it is by genes' (15). A story about hunting or about a place, according to Lopez, '. . . draws on relationships in the exterior landscape and projects them onto the interior landscape. The purpose of storytelling is to achieve harmony between the two landscapes . . .' (68).

Despite the individual and personal nature of most stories, and the various interpretations and meanings that one story may have for different members of a community, each person's story is interwoven with those of others to give a sense of continuity. Each person has their own footprints in the landscape, as it were, their own indelible memories that form the basis of stories that go to make up the community repertoire. Within a context of modern Greenlandic society, however, the cultural richness of place is often ignored by an official emphasis on a national Greenlandic identity, which is now coming into conflict with emerging local level interests and identities.

Since the introduction of Home Rule, modernisation policies have contributed to an increasingly industrialised and centralised Greenlandic Inuit society, marginalising some areas such as Upernavik district. Tensions and differences in environmental orientation are evident when the local significance of landscape in places such as Upernavik district is compared to the significance the environment has for the Greenlandic Home Rule government. Small community areas become lost in the larger, national map of Greenland. Local hunting grounds become Greenlandic territory and the environment is seen as something to develop (in terms of fishing grounds, and oil and mineral extraction) in the wider national interest. Greenlandic Home Rule is politically, economically and financially fragile and would be strengthened if economic needs were met by the commercial and intensive harvesting of both renewable and non-renewable resources. Dependence on foreign aid and knowledge would then be reduced. From this perspective the environment is becoming something with an explicit political and territorial meaning, invested with a degree of authority arrived at through legislation and agreement (Cohen 1976: 53–4).

In this way the diverse local perceptions and understanding of the environment to be found in the settlements in Upernavik are ignored. The study of place and locality, together with an understanding of the meaning of place names, illustrates the rich cultural fabric of modern Greenland. While there is a wider, national Greenlandic sense of identity (*kalaaliussuseq*), throughout Greenland there is also diversity in language, history and mode of production, ranging from industrial fishing to sheep farming and seal hunting. Local identity mediates national identity, but by emphasising nationalism and ignoring locality the heterogeneous nature of contemporary Greenlandic society is glossed over. The process of 'Greenlandisation', while an economic and political process, is above all ideological and finds expression through the dominance of West Greenlandic (Kitaamiutut) as the official language in Greenland (Nuttall 1990b). So, officially, West Greenlandic terms and names are now used for geographical features, and have displaced many pre-existing names that derived from the diversity of Inuit dialects to be found in Greenland.

Places in north-west Greenland, as indeed places throughout the world, are significant not because of any inherent value, but because of the meanings and cultural specificity attributed to them. In the words of Entrikin '. . . places become specific as we give them meaning in relation to our actions as individuals and as members of groups' (1991: 16). However, the study of Inuit perceptions of the environment and the mapping of hidden worlds as revealed through storytelling and memory has hardly begun, and it is to be hoped that there will be enough time to discover the cultural landscapes of the Greenlandic Inuit before modernisation and industrialisation marginalises subsistence hunting even further.

ACKNOWLEDGEMENTS

This chapter has arisen from a larger research project on changing political and cultural understandings of natural resources and the environment in Greenland. It draws on fieldwork conducted in north-west Greenland and funded by the Economic and Social Research Council (ESRC).

NOTES

1 Siumut is one of three main political parties in Greenland and has held power since the first elections in 1979. The other two main parties are the opposition Atassut, and Inuit Ataqatigiit.
2 Attachment to place, of course, is not the only basis for the cultural construction of identity in north-west Greenland. A fuller discussion of what Inuit identity is founded upon and given meaning by can be found in Nuttall (1992).
3 See Bird-David (1990) on hunter–gatherer notions of the giving environment.
4 Examples of the elaborate terminology used by Inuit to describe their physical environment can be found in Nelson (1969).
5 The period of Danish colonial rule was between 1721–1953,

when Greenland became an integral part of the Danish kingdom. Although Home Rule was granted in 1979, Greenland has remained a constituent part of the Danish Realm.
6 Devil's Thumb is a mountain that looks like a gigantic thumb when viewed from the south. The Inuit call it Kullorsuaq ('big thumb').

REFERENCES

Bird-David, N. (1990). The giving environment: another perspective on the economic system of gatherer–hunters. *Current Anthropology* 31: 183–96.

Basso, K. H. (1984). Stalking with stories: names, places and moral narratives among the Western Apache. In Bruner, E. (ed.) *Text, Play and Story*. Proceedings of the American Ethnological Society: Washington DC.

Cohen, E. (1976). Environmental orientations: a multi-dimensional approach to social ecology. *Current Anthropology* 17(1): 49–70.

Dahl, J. (1988). Self-government, land-claims and imagined Inuit communities. *Folk* 30: 73–84.

Entrikin, J. N. (1991). *The Betweenness of Place*. London: Macmillan.

Forchhammer, S. (1989). *Erhvervsøkonomisk og social udvikling i et Grønlandsk fangersamfund*. Unpublished Master's thesis, Institute for Eskimology, University of Copenhagen.

Freeman, M. (ed.) (1976) *Inuit Land Use and Occupancy Project*. Ottawa: Department of Indian and Northern Affairs.

Geertz, C. (1983) *Local Knowledge*. London: Harper Collins.

Gulløv, H. C. (1987). Dutch whaling and its influence on Eskimo culture in Greenland. In Hacquebord, L. and Vaughan, R. (eds) *Between Greenland and America*. University of Gröningen: Arctic Centre.

Haller, A. (1986). *The Spatial Organization of the Marine Hunting Culture in the Upernavik District, Greenland*. Universitet Bamberg: Bamberg.

Kleivan, I. (1986). De grønlandske stednavnes vidnesbyrd om vandringer og forskellige aktiviteter. In *Vort Sprog–Vor Kultur*. Proceedings of a symposium held at Ilisimatursarfik, Nuuk, Greenland 1981. Nuuk: Pilersuifik.

Kleivan, I. (1991). Greenland's national symbols. *North Atlantic Studies* 1(2): 4–16.

Lopez, B. (1989). *Crossing Open Ground*. New York: Vintage Books.

Nagel, T. (1986). *The View from Nowhere*. New York: Oxford University Press.
Nelson, R. K. (1969). *Hunters of the Northern Ice*. Chicago: University of Chicago Press.
Nelson, R. K. (1989). *The Island Within*. San Francisco: North Point Press.
Nuttall, M. (1990a). S. S. Triune: the loss of a Dundee whaler. *Polar Record* 26(158): 235.
—— (1990b). Greenlandic: political development of an Inuit language. *Polar Record* 26(159): 331–3.
—— (1991). Memoryscape: a sense of locality in northwest Greenland. *North Atlantic Studies* 1(2): 39–50.
—— (1992). *Arctic Homeland*. London: Belhaven Press.
Okri, B. (1991). *The Famished Road*. London: Jonathan Cape.
Petersen, R. (1985). Danske stednavne i Grønland. In Jørgensen, B. (ed.) *Stednavne i Brug*. København: C. A. Reitzels Forlag.
—— (1991). The role of research in the construction of Greenlandic identity. *North Atlantic Studies* 1(2): 17–22.
Smith, E. A. (1991). *Inujjuamiut Foraging Systems*. New York: Aldine de Gruyter.
Tuan, Y. F. (1975). Place: an experiential perspective. *The Geographical Review* 65(2): 151–65.
Vaughan, R. (1991). *Northwest Greenland: a history*. Maine: University of Maine Press.
Weeden, R. (1992). *Messages from Earth*. Fairbanks: University of Alaska Press.

GRAHAM HARVEY

Gods and Hedgehogs in the Greenwood: Contemporary Pagan Cosmologies

An increasing number of people today name themselves 'Pagan' and their spirituality 'Paganism'. They stress the association of the Latin-derived word *Pagan* with 'countryside' and 'country-dweller'. Other associations (such as the Christian ones, 'non-Christian' and 'one not enlisted in god's army') have significance to some Pagans. Some prefer the Northern European equivalent and name themselves 'Heathen'. Others prefer names indicating an affinity with a particular tradition, such as Druid or Witch. I hope I will not cause offence by using the name Pagan here. What makes it possible to use one name for the variety of groups, interests and people naming themselves Pagan is the centrality of the Earth or Nature in this spirituality.

My interest is in the ways in which Pagans see the world. The centrality of the Earth in their spirituality implies that the maps of their 'invisible world' ought to bear some relationship to the landscape mapped by the Ordnance Survey. However, whilst the Ordnance Survey marks the locations of stones, hills, springs, woods and paths, it does not show that these, and other, natural features have life and relationships. A map to accompany a walk through a nature reserve may represent the animals, plants and birds that can be encountered en route. A child's map is perhaps more likely to *animate* the trees and rocks as well as illustrating people and other living things. A Pagan celebrating the landscape might appear to be merely on a nature ramble or to be childishly animating her surroundings.

THE GREENWOOD

Though they believe that 'all the Earth is Sacred' Pagans are likely to have a special place or places to which they return often, especially for seasonal festivals. Pictures of the Earth taken from the Moon enable us to see our planet as one entity. Our culture makes connections between distant places (e.g. bananas, wars, TV and World Music), reinforcing the knowledge that the 'Earth is One'. However, it is not easy to convert that knowledge from an idea to an experience or a celebration. In a chosen 'special place', it is possible to experience and celebrate the Earth's life.

These special places can be named 'the Greenwood'. Once upon a time the whole land was covered with forest, only a small percentage of which now survives. In the remaining semi-natural woods, Pagans imaginatively enter a less human-centred world. It is a world where we are visitors and where we can encounter Life in all its diversity, in all its many dimensions and with all its relationships. The Greenwood can be entered not only in a wood: the special place of some Pagans is a single tree in a city park, encountered as the World Tree. We are also surrounded by our ancestors' artefacts: stone circles and cathedrals. These, too, can be places where the Greenwood is strong and alive.

GODS AND HEDGEHOGS

The Greenwood is a place full of life: plants, insects, animals, birds, fish and of course, the teeming bacteria of the soil. These 'other-than-human persons' (Hallowell 1960) have their own lives to live. They usually live without interest in our human lives, although we often disturb them, sometimes fatally. None of them is less significant than others, though some Pagans relate to trees but are only minimally interested in other plants and have no interest in insects. Others take into account the minerals on which the Greenwood depends.

The Greenwood is also inhabited by additional 'other-than-human persons' who are named variously according to the tradition or choice of the Pagan in the Greenwood. Some talk of 'Gods and Goddesses', others of 'Elementals', 'the Faeries' or other beings. There is a diversity of opinion on who these beings may be. Some say they are beings as real as humans and hedgehogs. Others believe this a poetic way of speaking about natural phenomena, personifying storms or growth, for example. Some may prefer to think of them as archetypes, existing 'inside' people (as individuals or groups) which may be evoked by a walk in a wood or in meditation at a chosen special place. Others are happy to be ambivalent in their attitude to these beings, addressing them as if 'out there' but believing that 'all deities reside in the human breast' (Blake 1793).

Gods and Hedgehogs in the Greenwood

PAGANS IN THE GREENWOOD

Having given a glimpse of the Greenwood with its other-than-human inhabitants, I shall now look at what Pagans do when they enter the Greenwood.

Attitudes (or beliefs) obviously cause different sorts of actions. I am interested here in those ('animists'?) who relate to 'other-than-human persons' who are as 'natural' as Pagans and Hedgehogs, though they may only be visible at special moments to rare human-persons (or even to rare hedgehog- or tree-persons).

Pagans might take walks in the woods to learn about the trees, plants and other 'wildlife'. Many Pagans read works of ecologists, biologists, herbalists and others to learn about the natural life of the woods. They might also study archaeology, folklore and mythology to learn about ancestral ways of relating to the divinities, faeries and other 'other-than-human persons' who might still be in the Greenwood. But such nature-rambles and study-trips are not the main activity of a Pagan in the Greenwood.

Pagans know themselves to be a small part of the life-cycle and relationships of the Greenwood (and of the whole Earth herself). They are also conscious of the ecological damage done by human lifestyles to the Earth and her inhabitants. When entering the Greenwood, Pagans enter a less human-centred world with respect, sometimes greeting 'other-than-human' guardians (significant trees, for example) at their chosen entry place.

Within the Greenwood Paganism can be described as 'sitting among the trees' rather than 'a spiritual path'. Other 'spiritual paths' may suggest that 'here' is not 'home' and 'I' am not 'holy' or 'enlightened'. Whilst there may be other paths going to interesting places, Paganism asserts that 'Earth is our home' and people only need to discover how to relate to and honour both themselves and others who surround them. Paganism celebrates the goodness of the Earth, the physical and the mundane: the 'here and now'.

At the seasonal festivals celebrated by Pagans, there will probably be fires burning, tended by individuals or groups. Some will engage in rituals, such as inviting 'other-than-human persons' to participate in some way. These will include the Guardians of the (4, 5 or 6) directions, ancestors and inhabitants of the place, such as gnomes or the faeries. The totem creatures of those present are likely to be there too ('imaginatively' and in 'mundane reality').

Sometimes it is traditional to cut parts of a plant, for example, holly and ivy at midwinter, hawthorn blossom at Beltain. Pagans will seek permission before cutting and may decide that the tree is saying 'no'. Some may make offerings to the tree to acknowledge that you cannot take without giving. Such offerings may consist of strips of cloth or strands of wool left in the tree or pieces of festival food and drink buried at its roots. A hug may be as effective a way of expressing a relationship with a tree. Protecting it from vandals may also be necessary.

In the heart of the Greenwood there might be a spring from which water can be drawn after suitable greetings, and around which gifts can be left. By way of celebration of the four elements, and as a gift, candles may be lit beside the spring, the light from the place joining the waters flowing from it in a celebration of the Greenwood.

All this activity should not obscure the fact that Pagans try not to be invaders and disturbers of the Greenwood. Even whilst sitting by the fire at festival times, maybe telling stories, they will be aware of the lives lived all around them. Human activity does, however, endanger the Greenwood. Tree felling or conversion to plantation will destroy a complex living environment and replace it with a simple human utility. Mythology says that the Old Gods withdraw further from each encroachment of humanity, Avalon disappeared into the Mists, the Faery into the Hollow Hills. Now they are only rarely met, and then only briefly and with some danger.

Pagans enter the Greenwood respectfully and when they return from their special place they do not abandon relationships and responsibilities gained there. Pagan spirituality is not about the worship of some greater power(s), still less is it about attempting to gain forgiveness and salvation (Tawhai 1988). It is about discovering (and enhancing) the relationships and fulfilling the responsibilities of sharing a small planet with a host of human and 'other-than-human persons'. If a Pagan rescues a hedgehog from a dangerous road this is not because the hedgehog was praying for salvation or will be eternally grateful. It is what must be done in redressing of the imbalance caused to 'other-than-human persons' by dangerous human lifestyles. Similarly, perhaps Pagans might meet the Faeries in the Greenwood, without prayer or praise taking place. There will be honour and awe, and in the encounter Pagans may hope to gain encouragement that their chosen special place is indeed the Greenwood. For the most part, Pagans, Gods and Hedgehogs will get on with their lives quietly aware of the web of life that is the Greenwood.

REFERENCES

Blake, W. (1793). Proverbs of Hell. In Keynes, G. (ed.) (1966) *The Complete Writings of William Blake*. London: Oxford University Press.

Bloom, W. (ed.) (1991). *The New Age: An Anthology of Essential Writings*. London: Rider.
Bloom, W. and Pogacnik, M. (1985). *Ley Lines and Ecology: an Introduction*. Glastonbury: Gothic Image.
Dames, M. (1982). *Mythic Ireland*. London: Thames and Hudson.
Eilberg-Schwartz, H. (1989). Witches of the West: Neo-paganism and Goddess Worship as Enlightenment Religions. *JFSR* 5: 77–95.
Graves, T. (1978). *Needles of Stone*. London: Thorsons.
Hallowell, A. I. (1960). Ojibwa Ontology, Behaviour and Worldview. In Diamond, S. (ed.) *Culture in History: Essays in Honour of Paul Radin*. New York: Columbia University Press.
Harvey, G. (1993). Avalon from the Mists: the Contemporary Teaching of Goddess Spirituality. *Religion Today* 8.2: 10–13.
Hutton, G. (1991). *The Pagan Religions of the Ancient British Isles: Their Nature and Legacy*. London: Basil Blackwell.
Maltwood, K. (1964). *A Guide to Glastonbury's Temple of the Stars*. London: Clarke.
Mathews, C. and J. (1985). *The Western Way: A Practical Guide to the Western Mystery Tradition*. London: Arkana.
Mitchell, J. and Rhone, C. (1991). *Twelve Tribe Nations and the Science of Enchanting the Landscape*. London: Thames and Hudson.
Roberts, A. (1977). *Atlantean Traditions in Ancient Britain*. London: Rider.
Tawhai, T. P. (1988). Maori Religion. In Sutherland, S. et al. *The World's Religions*. London: Routledge.

HILDA ELLIS DAVIDSON

Mythical Geography in the Edda Poems

One example of the desire to map invisible worlds can be found in the work of a group of unknown poets in Norway and Iceland, surviving mainly in one little manuscript book, the Codex Regius, known generally as the Elder or Poetic Edda. It was written about the year 1300, with prose notes added by one or more editors, and contains 11 poems about the world of the gods, together with a group of poems on legendary heroes.[1] Some of the mythological poems are thought to go back in their original form to the pre-Christian period, although the date and provenance of each is a matter of dispute. Mapping invisible worlds appears to have been a favourite occupation of these poets; as well as describing aspects of the Otherworld, they introduce a number of place names from the supernatural realms, and it is worth considering why they collected these names so assiduously. Was it a popular intellectual game among those versed in the lore of the supernatural world? Or an aid to memory for training young poets? Poetry was held in high regard in Iceland in the Viking Age, and its poets were honoured in many courts of northern Europe. Should we see in such names features of an invisible landscape, akin to that of the vision literature of the Middle Ages, with strong religious undertones, or perhaps linked with the interior journey of the shaman in earlier times? Or are they merely the product of learned men amusing themselves in the spirit of modern academics disputing over the private life of Sherlock Holmes? Another question worth asking is what type of landscape is attributed to the supernatural realms in the poems and tales which have come down to us.

Although we are told a good deal about the cosmology of the Viking Age, it is not easy to construct any clear map or diagram from the details given, and attempts to do so have not succeeded.[2] We gather that there were a number of separate worlds or regions inhabited by different races of beings, and the usual number given is nine. In the centre of these was the World Tree, with the different regions grouped around it, or possibly rising one above another, with some down below among the tree's roots. The tree was also the place where the gods met in their realm, which seems to be somewhere above the earth, the divine kingdom of Asgard with the halls of the gods within its sheltering walls. Here the Æsir dwelt, together with some of the Vanir, although the latter had a realm of their own called Vanaheim in the underworld, and it was from this that the goddesses came. Somewhere below, among the tree's roots, was Jotunheim, realm of the giants, and also Hel, the abode of the dead, with the realm of humankind, Midgard, the Middle Earth of Tolkien's *Lord of the Rings*, between them. But not even the ingenious Snorri Sturluson, much closer to the myths than we are, could produce a logical description of the cosmos in his *Prose Edda* of the thirteenth century. He throws us into confusion with his description of a bridge, which he tells us is the rainbow, running from heaven to earth, and then places the judgement seats of the Æsir under one of the roots of the World Tree, so that they ride out over the bridge to attend their court. I have found no convincing indication of the position of the bridge *Bifrǫst* (Quivering Road), sometimes called *Bilrǫst* (meaning uncertain). The second part of the name, *rǫst*, was used for a stage along a route, and would indicate a road rather than a bridge; De Vries (1957: 379) thought it possible that it originally meant the Milky Way extending across the night sky. The poems do not actually tell us that the bridge is the rainbow, and it may be a suggestion made by Snorri himself.

According to the poems, the central tree was an ash, called Yggdrasil; this name probably means Horse of Ygg, and Ygg was one of the many names of the god Odin, meaning 'terrible'. It was apparently a tree on which sacrifices hung, as we are told was the case with the great tree outside the temple at Uppsala (Adam of Bremen IV, 27: 208), and Odin himself is said to have hung on it in torment in order to win understanding of the runes which brought secret knowledge. It was perhaps from the top of the tree that Odin in his special seat could look out over all the worlds below. It is clear that Asgard was not an isolated kingdom, and that Jotunheim, land of the giants, lay somewhere beyond it. The giants made various attempts to reach it, and so there was always a guard at the bridge, for Asgard could be reached by an overland journey from Jotunheim. Thor journeyed regularly to Jotunheim in his wagon drawn by goats to slay members of the giant race and keep them in awe of his mighty hammer, and in one of the earliest poems (*Thórsdrápa*) he had to cross a raging river to reach it.

Mythical Geography in the Edda Poems   97

Sometimes Loki flew there in bird form, and it was evidently a long way from Asgard. As for the dead, there were differing suggestions as to where they dwelt. Sometimes they are said to be beneath the earth, passing in and out by way of a burial mound; sometimes they dwell in certain mountains, or are confined to their special realm of Hel, ruled over by a giantess of the same name, a sinister corpse-like figure called the daughter of Loki, who never had any dealings with other goddesses or giantesses.

One celebrated tale in the *Prose Edda* (*Gylfaginning* 49), which Snorri seems to have taken from a lost poem, tells of the ride from Asgard to Hel to rescue Balder after his death. The road is described as running through a dark and gloomy land, over rivers and mountains; it took the messenger Hermod (or is it Odin himself? – we are never sure) nine nights to reach the golden bridge over the Echoing River, guarded by a maiden. She tells him that a great company of the dead has recently passed that way in silence, but that he, who is not of the race of the dead, made a resounding clatter as he crossed, and she then directs him to ride down the road to the north to the gate of Hel. Sometimes again the dead are represented crossing to their destination by water, despatched, as was Balder himself, in a ship. Others, heroes of proved valour as well as kings and leaders, passed directly through the air to Odin's hall in Asgard, Valhalla, the abode of the slain. They were guided from the battlefield, where they fell at Odin's behest, by his attendant spirits, the valkyries. We know from the work of Franz Cumont (1949) how it was possible in the Roman world for seemingly contradictory ideas about the fate of the dead to exist side by side; they seem to express shifting imagery rather than intellectual beliefs and no static map could do justice to their variety and richness.

Indeed the cosmos presented in the Norse poems and stories can hardly be described as static. The impression is one of constant, restless movement; no race of beings remains at peace for long. The gods visit the giants and the underworld, and wander about the earth. Even the dead seldom lie quiet in the grave, but move in and out of their mounds and make the long journey to Hel's gate, or ride to Valhalla. The giants advance, if they can, on Asgard; a goddess is borne off to Jotunheim and has to be rescued by quick thinking and determined action, Loki's cunning supported by the might of Thor's hammer. Odin travels constantly between the worlds on sinister errands, riding his eight-legged steed or slipping into kings' halls in his broad-brimmed hat and cloak, and he bears such names as *Vegtamr*, Road-wise, and *Gangleri*, Wanderer. Whenever thunder is heard, it is said to be Thor's wagon on one of its noisy journeys across the sky, and he also travels on foot and wades over rivers. Freyr's golden boar, shared with his sister Freyja, races daily through the heavens and into the underworld, and their ship sails against wind and tide. There was little rest for the followers of the old gods, and the reward offered to great kings after death was an endless round of fighting and feasting, so that no swords were allowed to rust.

A memorable image of perpetual movement, destruction and renewal is that of the World Tree constantly devoured by the living creatures which feast on the branches while it puts forth new shoots to replenish them, a picture echoed in the restless, twisted carvings of the Viking Age with biting creatures constantly striving against one another. The outgoing energy of young Scandinavians in the Viking Age which drove them to travel thousands of miles, fighting, raiding and finding wealth in distant areas of Europe, and exploring new territory East and West, may be the inspiration behind this constant emphasis on movement and travel in the Otherworld. It is worth remembering, incidentally, that Viking sailors carried no maps or compasses. They found their way across oceans and up rivers by landmarks and the ability to recognise signs of land, together with knowledge of ocean currents and prevailing winds, much as the Polynesians in the more recent past found their way around the vast Pacific in their small boats (Lewis 1978). This partly explains why information about the Otherworld in the Eddas is not of the kind from which a map can be constructed, while at the same time it vividly conveys a sense of a vast landscape, long routes through a threatening countryside, and a series of landmarks.

The poems which provide us with our geographical information fall into three main classes. First, there are those concerned with contests in knowledge in the form of questions and answers exchanged between two supernatural beings, perhaps a god and a giant. There is much information in a poem known as *Vafþruðnismál*, Words of the Mighty Riddler, which is in the form of a dialogue between Odin, wisest of the gods, and an ancient giant famed for his knowledge, who can remember the beginning of the worlds. Many of the questions deal with cosmology; Odin is asked the name of the steed of the sun, the river dividing the realm which divides the land of the gods from the giants, and the field where the last battle will be fought. When Odin's turn comes to put the questions, he asks about the creation, the origin of the giants, and then about his own fate and what will come about after Ragnarok and the destruction of the gods. It was presumably because he needed to know such secrets that he sought out the giant, whom he finally overcame by a cheating question: what words did Odin whisper into the ear of Balder when his son lay on the funeral pyre? Then at last the giant realised the identity of his visitor, since Odin alone could give the answer to that riddle.

Another poem of this type is *Alvíssmál*, Words of All-Wise, in which the contest is between Thor and a dwarf who wants to marry his daughter. The questions here are all concerned with the different names used by various beings for features of the natural world, such as sun, moon, wind and sea. It has some relevance for a study of the cosmology, since it gives us a chance to complete the list of the Nine Worlds (Davidson 1975: 183). The worlds of humankind, of the two races of gods, Æsir and Vanir, of the giants and of the dead, are often mentioned, and the four remaining ones could be those of the Elves, the Dwarves, the heroes in Valhalla and a group called the High and Mighty Ones, possibly the most exalted gods. We have no authorised version, however, of these Nine Worlds, and they may always have varied in different regions and in the imagination of different poets. The impression left by the poems is that there was no special desire to strive after consistency, such as appeals to the modern mind, but that imagination might run free within the accepted structure of the Otherworld which the poets recognised.

A verbal contest of a slightly different kind is that in *Hárbarðsljóð*, Lay of Greybeard, a 'flyting' which takes place between Odin and Thor, consisting of taunts and boastings. Thor is on his way home to Asgard and Odin, disguised as a ferryman, refuses to take him over the water. It contains many references to lost myths concerning the exploits of the gods. Another flyting poem is *Lokasenna*, Loki's Mocking, where Loki makes his way into a feast from which the gods have tried to bar him, and insults individual gods and goddesses in turn, revealing old scandals and weaknesses. Here presumably it is the knowledge of exploits of the gods which is being tested.

The second group of poems is that purporting to be spoken by a prophetess or divine being with special knowledge of past or future. The long poem *Hávamál*, apparently several earlier poems pieced together, is represented as the utterance of the god Odin. Much of it consists of down to earth practical wisdom in which cosmology plays little part, but we have Odin's account of how he won the magic mead for the gods, and the famous passage which describes him hanging upon the World Tree in order to attain to understanding of the runic symbols. Another poem represented as the speech of Odin is *Grímnismál*, Words of Grimnir, the Masked or Hooded One, one of Odin's many names. This is an important source for our knowledge of Otherworld geography, supposed to be revealed by the god as he sat in torment between two fires. It describes the halls of Asgard, including Valhalla, and gives information about the World Tree, the journey of the sun through the heavens, and the various names of Odin.

The outstanding poem of this class however is *Vǫluspá*, Prophecy of the Seeress, a long poem thought to have been composed at the very end of the pre-Christian era, about the year 1000. It may indeed be by a Christian poet, but if so it was one well versed in the lore of the old gods and sympathetic to it. It is filled with a sense of loss, revealing how the world came into being, how the bright kingdom of the gods was established, and how the inevitable end came about with the giants' attack on Asgard, the breaking loose of the bound monsters, and the destruction of the earth by fire and the encroaching sea. It ends with a picture of the same earth rising cleansed from the waves for a new cycle of existence, under the rule of the young sons of the old gods, peopled by the descendants of two human beings who survived the destruction. This poem is filled with confident assumptions about the supernatural realms, as though these were familiar territory for the poet's audience. Poems of this type may have been popular, since a fragment known as the *Shorter Vǫluspá* is included in the Edda collection. Another poem giving an account of what is to come is *Baldrs Draumar*, Balder's Dreams, where a dead seeress is called up by Odin from her grave in Hel to reveal why Balder has been troubled by ominous dreams, and what catastrophe to the gods they portend. In *Hyndluljóð*, Lay of Hyndla, the goddess Freyja descends to the underworld with her golden boar, who is her young lover Ottar in disguise, in order to compel a giantess to recite the names of all his ancestors, so that he may learn his lineage. *Helreið Brynhildar*, Bryhild's Ride to Hel, is a different example of revelation. Brynhild, a character in the Sigurd cycle of heroic tales, journeys to Hel on a wagon after her voluntary death on a funeral pyre, and is challenged by a giantess, perhaps the same as the maiden at the Echoing Bridge encountered by Hermod, who questions her right to join Sigurd in the Otherworld. In reply, Brynhild relates what led to Sigurd's death and her own suicide. It may be significant that in these poems the speakers appear to yield their information reluctantly, after repeated and insistent questions from someone with the power to demand it. There is also an indication that one or both speakers may be disguised in some way, and only recognised when the revelation is complete.

The third group, in which the last three poems might also be placed, deals with journeys between the worlds, and the poems are partly or wholly in narrative form. In *Hymiskviða*, Ballad of Hymir, Thor visits the giant Hymir and on a fishing expedition hooks the Midgard Serpent, one of the bound monsters fated to break loose at Ragnarok. This was one of the most popular myths of the Viking Age, told in several early skaldic poems, and carved on a number of stones (Sørensen 1986). In *Thrymskviða*, Ballad of Thrym, a superb comic narrative poem, Thor's hammer, on which the safety of the gods depends, is stolen by a giant, and Loki flies far in bird form to find it. He accompanies Thor to Jotunheim with the god disguised as Freyja, ready to become the giant's bride, and so Thor recovers his hammer at the wedding feast. Another set of poems deals with the winning of a fair maiden, Gerd, a giant's daughter in the underworld. In *Skírnismál*, Words of Skirnir, the hero is either a messenger of the god Freyr or the god himself, and he makes a long and dangerous journey to the underworld to persuade Gerd to become the bride of the god. The journey is described through the dialogue; there are allusions to dew-wet fells to be crossed, to a wall of fire round the giant's hall, and to foes who bar the way. Similarly, in two other poems which are found in late paper manuscripts, *Grogaldr*, Spells of Groa, and *Svipdagsmál*, Words of Svipdag, the journey is again conveyed through the dialogue. In the first poem, Svipdag's dead mother from her gravemound recites to him the necessary spells which will enable him to make his way across mighty rivers, an ocean, and high mountains where the cold is intense, and to follow a misty road, where foes lie in wait, until finally he rides through the flames surrounding the hall where the maiden waits for him on the Hill of Healing. These poems help to build up a general picture of a long, wearisome journey over vast distances, and of hardship, danger and endurance, such as was all too familiar to the wandering Vikings in their progress over Europe and beyond. It is perhaps significant that such a journey is associated with a divine marriage, thought to have formed part of the cult of Freyr and the Vanir, fertility deities. In *Skírnismál* the final meeting takes place in a wood called Barri, which seems to represent a cornfield, and this is to save Gerd from complete sterility and annihilating death, which would be the result of the rejection of Frey's wooing (Olsen 1909). One might indeed view the comedy of *Thrymskviða* as the sacred marriage turned upside down; Freyja could never marry a frost-giant, for this would bring about destruction of the cosmos and of the goddess herself, and once Thor gets his hands on his hammer again at the wedding ceremony, the feast ends in death for the groom and the guests.

From these three groups of poems we get the impression of the Otherworld divided into separate realms, but with plenty of opportunity to pass from one to the other, and the world of humankind only one among nine. We are led to think of roads, tracks and waterways occupied by many travellers, moving in ships, on horseback, by wagons and sledges, or on foot. Such a picture, incidentally, is borne out by many travelling figures on foot or in vehicles shown on a ninth century tapestry recovered from the Oseberg ship burial in southern Norway, which appears to show supernatural characters in the restored section (Krafft 1956). This insistence on the roads and rivers of the Otherworld might imply that it was important for men as well as the gods to possess knowledge of entry, and of routes to take when travelling to the land of the dead or down into the underworld in search of wisdom.

The place names in the poems belonging to this vast and mysterious world may be put into several groups. First, we have the names of the dwellings of the gods, from which we can learn something of their divine nature, even though many deities mentioned had been largely forgotten by the Viking Age. Thus the mysterious Ull, once perhaps a major sky-god, has a hall called Yewdale, supporting the theory that the World Tree may originally not have been an ash, but a yew. Some names are in accordance with what we know of the great gods of the Viking Age: the hall of Njord, one of the Vanir, is Enclosure of Ships, *Nóatun*, emphasising the importance of the ship symbol, and perhaps of ship funeral, in the Vanir cult. The conception of Asgard up above in the sky is strengthened by names like Hill of Heaven for the hall of Heimdall, and Broadgleam for the home of Balder.

Secondly, there are many names of rivers, harder to explain. In *Grímnismál* the slain wade across the river Thund, meaning 'raging' or 'thundering', presumably on their way to Hel, and this may be the reason for the tradition found in *Vǫluspá* and again in the Danish historian Saxo that there is a river in the underworld which carries weapons in its current; the name in *Vǫluspá* is *Slíðr*, 'Terrible'. The dead, though not presumably the distinguished ones riding to Valhalla, might perhaps be thought to drop their weapons as they waded over. The idea of a river to be crossed by the dead is an image familiar to us all: Charon acted as ferryman to the Roman dead, Mr Valiant-for-Truth and Christian struggled through deep waters to reach the Celestial City, and the Negro spirituals remind us of that one more river, the River of Jordan, that awaits us all. Another explanation for weapons in the flood is that they were some form of punishment threatening legs of the uncharitable dead who waded across, like the thorns in the Lykewake Dirge (Aubrey 1972: 31ff.), but I doubt if such a judgemental idea would appeal to the earlier Scandinavians. Certainly the importance of a river as a boundary between the worlds is well established; in the poems, the river Ifing (meaning not certain) lies between the realms of giants and gods, and is said never to freeze over, presumably because the influence of the gods proved stronger than that of the frost giants.

The interest in river names reaches its height in *Grímnismál*, where as many as 36 river names are given, most of them said to flow from the horns of the hart which eats the branches of the World Tree. The river names are spelt in various ways in different manuscripts, and the list is generally assumed to be an interpolation, possibly by some learned editor. One of the names is recognisable as that of the Rhine, well-known from the heroic Edda poems, and another is thought to be the Dvina, a river of great importance for Vikings going east. Christopher Hale (1983) concludes that many of the names may have been those of actual rivers in Norway, since they belong to an early type of name formation, making it unlikely that they have been invented by the poet. Rivers were a major factor to take into account for the Vikings journeying into Europe, and particularly into the dangerous eastern regions; it was important to know if they could use ships or boats on them, whether portage would be necessary (as on the Dnieper going to Constantinople), whether they might be defended at key points, and where ambushes could be expected. Thus it is hardly surprising that they play a major part in the dangerous journey into the Otherworld.

Bridges too were important. There is the Echoing Bridge leading into Hel, strangely said to be of gold, lighting up the darkness of the last lonely road. The Echoing River, which it spans, is mentioned again in *Grímnismál*. The bridge Bifrǫst, a means of entry into Asgard, has to be guarded against the giants, and Heimdall remains on guard there with his horn to give warning of the approach of the enemy. On the last day the sons of Muspell, who ever they may be, will ride on to it and it will collapse under their weight, and the final battle will begin.

The naming of mountains is not very common in the Otherworld accounts, but there is the wood Barri, thought to signify a cornfield, where Freyr claims Gerd as his bride, and another fearsome forest called Ironwood (*Jarnviðr*) to the east of Midgard, the abode of an ancient giantess whose sons were in wolf-shape, among them the terrible Fenriswolf who was ultimately to break from his bonds and devour Odin. These last names, and also

Bifrǫst, come from the important group dealing with the lore of Ragnarok, the last great battle of the gods. As to where it will be fought, the poets do not always agree; in one poem it is *Vígríðr* (Warpath) and in *Fafnismál*, one of the heroic poems of the Edda, it is the island *Óskópnir* (not made), a name which may indicate that the site is not yet made ready. The hall *Gimlé*, said to survive the destruction, means something like 'fire-refuge', and there presumably the sons of the gods take shelter. There must have been a whole wealth of lost lore in the Viking Age concerning Ragnarok, and many of the journeys taken by the gods were either to prevent the coming of the end, or to acquire more knowledge about this terrible last day.

It seems plausible that one reason for collecting and multiplying names was as an aid to memory, and no doubt these verbal contests in the Edda poems mirror similar ones between poets and story-tellers in the familiar world. It is not unlikely that something akin to our Brain of Britain and other types of radio and TV quiz games already existed in the Viking Age. One method of jogging the memory and perhaps teaching the young was pointed out many years ago by Magnus Olsen, who discussed some rhythmical lists from Norway from the thirteenth century onwards recording various places along a boundary, including that between Norway and Sweden (Olsen 1931: 153). Such lists of names are also known from Iceland, giving a series of place names and sometimes the route from one to the next is a kind of chant, easy to remember, such as '. . . from there to Langsio, from Langsio to Flysbek, from there to Holy Thorn, from Holy Thorn to the grey stone' and so on. Olsen suggested that Odin's reply to Thor's question as to the way home by road in *Hárbarðsljóð* might echo instructions of this type:

. . . Some way to the stock, then some way to the stone,
Follow then the left hand path
Till you come to the land of men.

In his analysis of river names, Hale suggested that scraps of lore of various kinds found their way in course of time into the mythological poems of the Edda and became mixed with that concerning the world of supernatural beings, thus increasing its complexity.

It seems that there is no simple explanation of the Otherworld place names in the Edda poems. Some no doubt originated in oral tradition, while others are the result of learned editing, and copyists' errors; poets' ingenuity no doubt also played a part. But there are some clues to possible reasons for the names, apart from providing an aid to memory. They tend on the whole to be associated with divination and the gaining of hidden knowledge, offering power to those who possessed it. They have links with the sacred marriage which renews the earth, possibly because this necessitates a journey to the underworld, and are linked also with the destruction of the world and fall of the gods at Ragnarok. The multiplicity of such names, and the value apparently put upon knowledge of them, might be partly due to the keen interest in travel in strange regions such as the forest lands of eastern Europe and the mountains of the Caucasus; such journeys promised rich rewards and a chance to see great marvels (Davidson 1976: chapters 3, 6). Exploring dangerous areas of the present world in the Viking Age demanded courage and endurance, and therefore provided an appropriate pattern for adventures in the Otherworld. How far there may be memories of the interior voyages of the shaman in such traditions is hard to determine, but there are many echoes of shamanic imagery in Eddic poetry, particularly that which has to do with Odin, conductor of the dead into the realms beyond. Certainly there can be few people more likely to be fascinated by the charting of the unknown than the Scandinavians of the Viking Age, and it is hardly surprising if it has influenced the cosmology of the Edda poems.

NOTES

1 A good edition of these poems is that edited by G. Neckel, *Edda, die Lieder des Codex Regius* (4th ed. 1962, rev. H. Kuhn). A readable translation is by Olive Bray, *The Elder or Poetic Edda*, Part I, Viking Club, 1908. That by H. A. Bellows, *The Poetic Edda*, American–Scandinavian Foundations, London 1926, is accurate but awkward. Good translations of some poems can be found in B. Phillpotts, *Edda and Saga*, Home University Library, London 1931, together with helpful discussion.
2 The cosmology is described in H. R. E. Davidson, *Gods and Myths of Northern Europe*, Penguin Books 1964 (chapter 8), and in *Myths and Symbols in Pagan Europe*, Manchester University Press 1988, pp. 167 ff.

REFERENCES

Adam of Bremen. Tschan, F. J. (trans.) (1959). *History of the Archbishops of Hamburg–Bremen*. New York: Columbia University Press.
Aubrey, J. (1972). Remains of Gentilisme and Judaisme. In Buchanan-Brown, J. (ed.) John Aubrey, *Three Prose Works*. London: Centaur Classics.
Cumont, F. (1949). *Lux Perpetua*. Paris: Géuthner.
Davidson, H. R. E. (1975). Scandinavian Cosmology. In

Blacker, C. and Loewe, M. (eds) *Ancient Cosmologies*. London: Allen and Unwin.
—— (1976). *The Viking Road to Byzantium*. London: Allen and Unwin.
Hale, C. (1983). The River Names in *Grimnismal*, pp. 27–39. In Glendinning, R. J. and Bessason, H. (eds) *Edda: a Collection of Essays*. Manitoba: University of Manitoba Press.
Krafft, S. (1956). *Pictorial Weaving from the Viking Age*. Oslo: Dreyers Forlag.
Lewis, D. (1978). *The Voyaging Stars*. Fontana Books: London.
Olsen, M. (1909). Fra Gammelnorsk Myte og Cultus. *Maal og Minne*, pp. 17–36. Oslo.
—— (1931). Deildevers. *Maal og Minne*, pp. 150–53. Oslo.
Sørensen, M. (1986). Thor's Fishing Expedition. In Steinsland, Gro (ed.) *Words and Objects: Towards a Dialogue Between Archaeology and the History of Religion*, pp. 258–78.
Vries, J. de (1957). *Altgermanische Religiongeschicte*. Berlin: de Gruyter.

URSZULA SZULAKOWSKA

The Pseudo-Lullian Origins of George Ripley's Maps and Routes as developed by Michael Maier

The English alchemist Canon Sir George Ripley of Bridlington created an analogy between a sea voyage and the alchemical process in his poem *The Compound of Alchemy* or *Liber Duodecim Portarum (The Book of the Twelve Gates)* (Mangetus 1702: vol. 2, 275–85).[1] Written in the late fifteenth century, the poem describes a castle with twelve gates. Each is named after one of the processes of alchemy and instructs in a certain alchemical procedure.[2] Ripley's journey, however, is not in a straight line for the castle is circular and the gates stand in the circumference of the walls. Furthermore, at several of the gates,[3] and, in particular, numbers two, 'Solution', and ten, 'Exaltation', there is the suggestion that entry of each gate constitutes a type of 'voyage' by sea. Ripley does not specifically state this but his phraseology implies such an image. He speaks of 'turning the wheel' to take a course into distant corners of the earth. Thus, he visualises the alchemist as a helmsman on a ship. By his 'wheel' Ripley is referring to the 'wheel of the elements': the medieval theory, following Aristotle, of the circulation of the four elements by which motion they transmute themselves into each other.

The present study suggests that Ripley's image of voyaging was elaborated by the seventeenth century alchemist Michael Maier in two alchemical treatises: the *Viatorium, hoc est, De montibus planetarum septem sue metallorum* (1618) and in his famous illustrated *Atalanta Fugiens* of the same year (Yates 1986: 87–8). Maier's imagery has not been placed by historians within the context of an original influence from fifteenth century pseudo-Lullian alchemy on his own source in Ripley. (Though Maier was not directly, nor consciously, influenced by pseudo-Lullian alchemy.) The intention of this chapter is to examine this historical line of iconographical influence from the early fifteenth to the early seventeenth century: from Italy to England to Prague. It raises questions concerning the meaning and purpose of abstract and naturalistic illustration in the Renaissance Hermetic tradition. In addition, a study of the voyaging theme reveals the elemental philosophy of alchemy, a materialistic monism which declares that all is one matter, both body and soul.

Jung, in his *Psychology and Alchemy* (1980), reviewed the theme of the alchemical 'peregrinatio'. Like the psychologist Silberer earlier (1971 [1917]: 233–335), Jung (1980: 369ff) interpreted the alchemical journey as a psychological process unifying and balancing conflicts within the individual. Whatever the explanation may be in modern psychological terms, however, in historical origins the travels of the alchemists probably derive from shamanic visions of the soul's journey to the heavens and to the underworld. This was an initiation which granted power, liberation, immortality and enlightenment and was regarded as a second birth or even as a reincarnation in a new body and persona: one may describe the adept as being 'transmuted' from lower to higher being (Eliade 1977). Conversely, alchemical matter in the process of transmutation inherits the anthropomorphic character of the shaman. Nearer in time, an immediate contemporary exemplar could have been the knightly quest of medieval chivalric myths. In Maier's case, Renaissance voyages of discovery influenced his thinking, for he begins his journey in Europe, thence to America, to Asia and finally to Africa. He moves with the sun, circumnavigating the spherical globe of the seventeenth century earth. Another prototype for him was the Greek myth of the voyage of the Argonauts to find the Golden Fleece. The notion of alchemical journeying reaches its full development in the Rosicrucian texts of the seventeenth century. Jung placed Ripley and Maier side by side amongst many others in his discussion but he was not concerned with the possibility of actual historical influences. Jung rightly identified Ripley's *Duodecim Portae* as a circular 'peregrinatio' but did not provide a closer analysis (1980: 381).

The Second Gate of Ripley's castle instructs the alchemist on the process of Solution. This involves distillation in which matter is purified by circulation in the flask in order to draw off the dross. It is the central process in alchemy. Ripley tells us that we must 'turn our wheel' according to 'altitude, latitude and profundity'. We must make our entrance in the west and then take our passage into the north. There, all our lights shall go out and we shall have to abide for ninety nights in the darkness of purgatory. Thence we must make our course up to the east. (In medieval maps, east was placed at the top of the map). Here we shall see many different colours appear. Thus both winter and spring will pass. The journey to the east is described as an ascending towards the sun which rises up with the daylight. Here, too, we shall spend our summer delightfully and our work will become perfectly white.

From the east, we descend into the south where we are told to set up our 'chariot of fire' and there shall be harvest, the end of all our work, fulfilling all our desires. Here the sun reigns in his own sphere and is red with glory after his earlier eclipse. He is the king reigning over all metals and mercury. Finally, we are told that all this must be done in one glass which has the shape of an egg and is well closed (Mangetus 1702: vol. 2, 277ff).

At this point, Ripley describes the alchemical procedure in its entirety as the four traditional colour stages of alchemy: the black ('nigredo' which he describes as 'purgatory'); the coloured stage (often called 'cauda pavonis' or the 'peacock's tail' but here described as spring with its flowers); and the white (the 'albedo' or final purification of base matter, also personified as the 'white queen' or the silver-producing stone). The final red is the gold-making philosopher's stone itself, which Ripley personifies as the sun–king reigning 'in all his red glory after his travails in the underworld of death' (his 'eclipse') (Coudert 1980: 42–3).

The distillation procedure in alchemy was extremely prolonged (Sherwood Taylor 1976: 72–3). Alchemists from the earliest days in the second century AD, probably in Hellenistic Egypt, saw this process in terms of a descent into the underworld and an ascent to the heavens (Sherwood Taylor 1976: 40–6). Ripley, however, modified this vision into that of an earthly journey across the plane of the world itself (which is clearly a flat disc for Ripley). He turned eschatological myth into a traveller's journal in the prosaic everyday world. This itself has a significance which will be further investigated within the present argument.

Thus, Maier's fictional journey in the *Viatorium* may have used as its model the autobiographies of Renaissance alchemists which describe their travels in quest of the stone (Doberer 1979: passim). Ripley had himself provided an account of his own journeys in the introduction to the *Duodecim Portae* (Mangetus 1702: vol. 2, 275ff). Maier later translated the treatise into Latin and published it in his *Symbola Aureae Duodecim Nationum* in 1617, the year before he published the *Viatorium*.

But, generally, Ripley's imagery of the voyage fits within the ruling metaphor of the alchemists, i.e. the comparison between the macrocosm of the world and the microcosm of the alchemical glass (Sherwood Taylor 1976: 115). Ripley's reference to the four alchemical compass points is an image of the four elements of Greek and medieval European science, namely, earth, water, fire and air.

The alchemists always claimed to follow Nature under God's guidance. They simply enabled Nature to act out her usual processes within the alchemical flask at a much faster pace. 'Natura naturans' was their motto. And Nature's aim was the perfection of all creation to become a reflection of its divine creator. Furthermore, God's own spirit was the soul of matter which moved through Nature. It had to be freed from its imprisonment in imperfect substance by the redemption of gross matter (Pagel 1961: 117ff).

... Gaudet natura naturam. natura naturam vincit ... (...)
... natura sequitur naturam, et amplectitur et sic stat opus philosophorum perficitur (sic) (Florence, Biblioteca Laurenziana Ms Ashburnham 1166, f.9r)

In short, the alchemists wanted to create matter's perfect body united to its pure soul: to unite Nature and God (Sherwood Taylor 1976: 114). They had three fundamental concerns: chemical work on matter; spiritual work by the alchemists on themselves; semantic work to find appropriate language and emblems, both verbal and visual, which could express the union and even the identity of spirit with matter. Their practical aim was the transmutation of base matter into gold through the creation of the philosophers' stone.

Profoundly influenced by late Renaissance English Hermeticism and alchemy (Yates 1986: 81–2, 87–9), Maier at the court of Prague elaborated Ripley's simple scheme into the description of a journey to the corners of the earth. Maier was inspired by the elaborate cosmological diagrams of the spiritual alchemist Robert Fludd in seventeenth century England (Klossowski de Rola 1988: passim). Most other written texts, in comparison, only give a general indication of alchemical terrain. In fact, it is the visual illustration which is responsible for the detailed treatment of geographical siting. It is, however, usually imagined as a shorter walk through a courtly garden attached to a castle rather than a sea-voyage. Herbert Silberer analysed such a brief but painful scenic route in a then little-known Rosicrucian treatise (Silberer 1971: 1–14).

Seventeenth-century illustrators developed a standard visual repertoire under the influence of contemporary landscape and townscape paintings and engravings of the Low Countries. Such topographies can be traced back to manuscript illustrations of the late fifteenth century, or specifically to alchemical treatises such as

The Pseudo-Lullian Origins of Ripley's Maps   111

the *Splendor Solis* (London, British Library Ms Harley 3469). This Venetian alchemical manuscript of 1582, written by Salomon Trismosin, is lavishly illustrated with realistic views of palaces, towns, fields, rivers and caves. Thus, by the seventeenth century, alchemical engravings showed coastlines, town squares, enclosed gardens, open ploughed fields and high mountains. The basic concept expressed is that of enclosure (gardens, city-scapes) versus the open countryside. The images are a visual expression of the alchemical concept derived from the Greeks of form (the perfect order of God) versus chaos (impure prime matter) (Sheppard 1960: 98–110).

Maier owed many fundamental ideas to the much-travelled Elizabethan magus John Dee (French 1987), who had reformed alchemy and practical magic in English Protestant intellectual thinking by reinterpreting Hermeticism (Clulee 1988) and was an intermediary influence between Ripley and Maier. It is well-established that Dee owned several copies of Ripley's *Duodecim Portae* and of the *Ripley Scrowles* in his (effectively) public library at Greenwich (Roberts and Watson 1990). He made notes and emendations on Ripley's texts and wrote a preface to the *Liber Duodecim Portarum* which was first published in London in 1591 in English as the *Compound of Alchymy* (French 1987: 82). It has already been mentioned that Maier, in fact, himself had published his own Latin translation of the *Duodecim Portae*. At a later date, Elias Ashmole in his *Theatrum Chemicum Britannicum* of 1652 (Preface) mentioned Dee in connection with Maier and stated that Maier collected, copied and translated English texts. Ashmole included the verses of Ripley's *Scrowle* and of the *Compound of Alchemy* in his compendium of English alchemy, *Theatrum Chemicum Britannicum* (1652: 375–9). The *Duodecim Portae* and another of Ripley's texts, the *Cantilena Riplaei* (Sherwood Taylor 1946: 177–82), also appeared in an *Opera Omnia* of his writings in Kassel in 1649. The general regard for Ripley's work was, thus, due to both Dee's and Maier's promotion in Europe. Copies of the *Scrowle*, for example, state that it was painted in Lubeck in 1588 (Klossowski de Rola 1973: Plate 65) and this fact seems related to Dee's extensive travels over Europe. One of Maier's dominant images is his quest for the Phoenix as an emblem of the philosophers' stone, i.e. of perfection rising from the purified ashes of its old self. Although not unique to Ripley it is the central visual image of his *Scrowle* and it may have been the source for Maier's image (*Symbola aureae mensae duodecim nationum* 1617: 572ff).

The historical existence of George Ripley (at work between

1450 and his death in 1490) as a cleric of rank and distinction is well documented (Brann 1979: 212-9). His various treatises on alchemy exist both in manuscript and in later printed forms (Linden 1990: passim). Some manuscripts purport to be of fifteenth century origin, self-dating at 1470 (British Library Ms Sloane 3667, f.52). However, an accurate chronology of the texts purporting to be Ripley's has yet to be produced. Nonetheless, there is a secure consistency in the theory of the texts, in particular, of influences from pseudo-Lullian alchemy (chiefly the *Testament*). The texts so argue internally for their authenticity that no authority has yet disputed their attribution to Ripley (Thorndike 1953: vol. 4, 349-52).

Pseudo-Lullism is a complicated series of alchemical treatises produced by anonymous authors from the late fourteenth to the eighteenth centuries, initially in Catalonia and in Northern Italy. They were spuriously attributed to the thirteenth century Catalonian mystic Ramon Lull (Pereira 1989: 1-5). Professor Percira has argued for the influence of late fourteenth and early fifteenth century pseudo-Lullists on English alchemy (1989: 22-5).

Ripley himself states in his preface to the *Duodecim Portae* that he learned his alchemy in Italy (Mangetus 1702: 275ff). Thus, Ripley's notion of the alchemical journey as a circulation through the four compass points may well derive from the diagrams of pseudo-Lullian alchemy. The conceptual foundation of pseudo-Lullian circles is a numerology of four, derived from the number of the material elements. Their visual format and mechanical mobility is founded on Aristotle's square of the elements. Ripley turned the diagrams into a 'map' which depicted the circular routes followed by the substances in the alchemical flask.

Pseudo-Lullism's and Ripley's elemental voyages are a disquisition on the nature of Being itself, that is, on the relationship of spirit to matter. Such was the alchemists' real concern. The dualistic Gnostic myth of the descent of the soul into its eternal opposite, matter, and of the soul's redemption by a suffering hero is an undoubted source (Sheppard 1960: 98-110). The Babylonian origins of this may lie in the Marduk/Tiamat myth. Alchemy is a depository of wisdom and debris from all possible European and Near Eastern philosophical and religious sources.

Much of the alchemists' thinking about materiality originates with two types of Hellenistic Gnosticism. The better-known are the dualistic beliefs of Iranian origin concerning the absolute distinction between two equal, inimicable principles, spirit and matter (Sheppard 1957-8: 86-101). There is also a monistic Gnostic influence in alchemy derived from the Sabbeans (Harranites) of the eighth and ninth centuries AD in the Near East (Sheppard 1962: 83ff). These viewed spirit and matter as being a single divine substance (Plass 1982: 69–72). The monistic strain in alchemy, according to Professor Pagel, was a significant influence on Paracelsus in the sixteenth century (1961: 117ff). Historians' usage of the term 'pantheism' in relation to alchemical theory is vague. Yet, it is important to understand the ontological distinctions in order to comprehend the purpose of the alchemists, in so far as that is at all possible (Szulakowska 1988).

There exists, for example, a Manichean ancestor of the alchemical myth of the redemption of the soul through a journey to salvation. This is the beautiful Syrian *Song of the Pearl* of the second century AD. In this well-known poem, a divine Prince descends into Egypt, the land of Hermeticism and of alchemy (Widengren 1965: passim). Commanded by his heavenly parents, his mission is to recover a precious Pearl guarded by a dragon. The prince is seduced by Egypt's materialism and loses his memory until woken to his true identity by a letter/bird sent by his parents. He subsequently fights the dragon and rescues the Pearl. Then he is united with his divine origins. In the *Song of the Pearl* are found all the elements which are the foundation of alchemical imagery: the hero prince (alchemist), sent forth on a mission by higher powers, to find the philosophers' stone (pearl), which is trapped in a lowly and despicable place (the anima of the metals trapped in gross corrupted matter). It is guarded by a venomous dragon which must be defeated. The dragon later becomes alchemical mercury and an image of the distillation process. In contradictory terms, the dragon can represent both impure venomous matter, as well as a pure spirit guide, an ambiguous and ambivalent go-between for heaven and earth. For, he is both the fallen redeemer and, yet, the very life blood (menstruum) of the prima materia.

Memories of the Manichean story are found again in a written description by the sixteenth century pseudo-Albert of an illustration originally found in a pseudo-Lullian manuscript, Florence, Biblioteca Nazionale–Centrale [BN] II iii 27, f. 112v. It was painted by Gerolamo da Cremona in *c.* 1470. Albert describes the dragon winding its way underground, as well as flying through the cosmic spheres of the heavens. It can be killed only by a hero, poor and naked (and wearing an anxious expression in Gerolamo's illustration). He alone has the power to kill the venomous dragon that the rich and powerful of the world fear (Albertus Magnus (pseudo) *Alexander Super arborem Aristotelis, Theatrum Chemicum* II.527). Pseudo-Albert states that Alexander the Great encountered this scenario in the course of his travels. In actuality, the real author of this treatise appears to have himself found the illustration in a Venetian copy of the Florentine original and recreated the ancient story around it (Szulakowska 1986: 53–72). The original illustration in the pseudo-Lullian text of Ms BN II iii 27 is well-publicised (Purce 1974: Plate 5). It shows a round fountain-basin set in the midst of the circular spheres of the seven planets or metals. The dragon winds down a young tree on which hang the diseased heads of the metals in need of purification and healing. There exists an illustration of the sixteenth century in one of the Ripley *Scrowles* in the British Library (see Figure 1) which recalls the Florentine image so suggestively that BN II iii 27 or its Venetian copy seems to be its source. In the *Scrowle* the fountain has been turned into a type of 'castle' (Jung 1980: Fig. 257). Perhaps this has occurred as a result of the influence of this image on the *Duodecim Portae*, although the respective dating of the two works has not been determined by scholarship. However, in the *Scrowle* are shown seven turrets on the circumference of the castle-fountain representing both the seven planets and the seven stages of the process, whereas in the *Duodecim Portae* the castle has twelve gates and processes. Thus, the numerology does not correspond. The fountain may represent the alchemical flask in which the conjunction and dissolution of male and female principles takes place, whereas the castle may be a symbol for the alchemical furnace. Thus, the discrepancy in the numbers of 'pinnacles' and 'entrances' may be related to differing procedures. There exist many illustrations of circular alchemical furnaces some of which take the forms of turreted castles (Florence, Biblioteca Laurenziana 1166, ff.18v, 19v, 21r, 23r; see Figure 2 and Carbonelli 1925: 111, 112, 113, 115).

The ancestry of the concept of circularity in alchemy is found in Orphic and Pythagorean myths of reincarnation and immortality. Orphic concepts of the immortality of the soul are probably of Vedic Indian origin. The Greeks created a radically different shape of time–space from the Judaic–Christian. For they constructed their cosmology and, indeed, their map of time, as a circle which could encompass the necessary reincarnation of the soul (Cornford 1980: 160–204). Plato's spindle of Necessity in Er's vision is a flat disc set on the vertical cosmic axle (*Republic*: 616Bff). Due to this circularity of time–space, the cosmos was believed to be eternal, reincarnating itself at the end of the cycle of the Great Year (Cornford 1980: 178–81). This idea was argued by Greek Aristoteleians and neoplatonists both. As far as the alchemists were concerned, though the texts are profuse in their Christian devotion, the circularity of time was a fundamental part

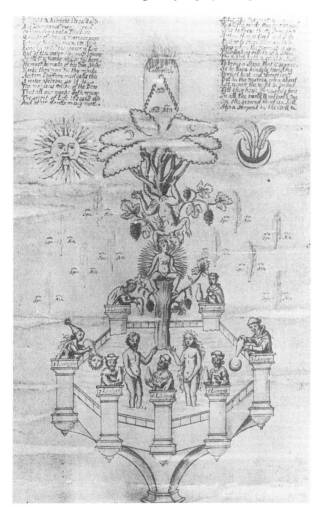

Figure 1　George Ripley, *Scrowle*, British Library Ms Add. 5025, Roll 3, dated 1588.

of their theoretical structure. For transmutation from lower to higher being was effected through a process of circulation and distillation. Furthermore, the perfect stone was created by returning base matter to its original womb, whence it was reborn as a pure reincarnation of its formerly corrupted self.

Distillation was invented by Hellenistic alchemists (probably by Zosimos *c*. 300 AD) and developed by the Arabs (Sherwood Taylor

Figure 2 Florence, Biblioteca Mediceo-Laurenziana Ms Ashburnham 1166, f. 21r, *Alchemical furnace* (*c.* 1470).

1976: 40–6), but the physical theory is Aristotle's (*De generatione et corruptione*; Multhauf 1956: 329, 346). Aristotle demonstrated that the changing nature of the world is the result of the transmutation of the four elements into each other (*On Coming-to-be and Passing Away*; Joachim 1922). This occurs through a mean. Since each element had two qualities (fire had heat and dryness), opposites could unite by moving through their common quality (Pagel and Winder 1974: 93ff). For example, fire could turn into its opposite, water, by passing first through one of the qualities of either earth or air. It could move through the quality of dryness shared with earth and then could move through the quality of coldness earth shared with water. Thereby fire and water were united. This was the major alchemical conjunction, also pictured as the marriage of sun and moon, sulphur and mercury, spirit and body. These would be the 'parents' of the hermaphroditic child, the stone of the philosophers.

In the middle ages, Aristotle's theory was pictured as a square with the elements placed at the four corners, the opposites being on the diagonals to each other (see Figure 3). Around the edges of the square the interchanges and mutations were effected in a circular movement (Yates 1954: 149, Fig. 2). Hence, the symbolic picture of the alchemical journey was not just that of passing around the circumference of a circle, but of *squaring* the circle. Ripley's round castle with its four compass points is an image of Aristotle's elemental theory. This geometrically impossible feat becomes especially important in the mysticism of the Rosicrucian alchemists. Maier, attracted by Dee's concepts, wrote his *De circulo physico quadrato* (Oppenheim 1616) on the theme. It is illustrated in his *Atalanta Fugiens* as an alchemist squaring a circle with dividers (*Emblem* XLVI; Fabricius 1976: Fig. 369). The concept is monistic. Nature (the four elements symbolised by the square) becomes God (symbolised by the perfect form of the circle).

There are three subtle ideas in the above theory which deeply influenced the alchemy of the pseudo-Lullians and, thereby, Ripley. These are the concepts of circularity, motion and of the medium between opposites. In pseudo-Lullian alchemy, there is an especially pronounced emphasis on movement and on the necessity of the mean or medium (which Ripley tells us is mercury), much more so than in other alchemical texts of other schools (Thorndike 1953: vol. 4, 3–64). Pseudo-Lullian treatises and those of their followers such as Ripley are immediately recognisable as a discrete group owing to the constant reiteration of these three principles. Thus, in Ripley's *Duodecim Portae* there

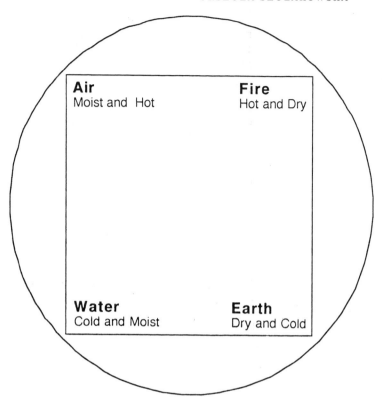

Figure 3   The elemental square of Aristotle.

are explanations such as the following in the Tenth Gate of Exaltation, namely that fire is in water. By turning the wheel of the elements, air should be converted into its opposite, earth, for air is in water which is in earth. Then water must be turned into its contrary, fire, for earth is in fire which is in air. We are instructed to begin the circulation in the west and go forth to the south where the elements will be exalted. Finally, Ripley summarises his instructions by stating that it is impossible to go from one extreme to another save by a mean because you cannot join together qualities contrary to each other by a direct route (Mangetus 1702: 280ff).

The prototype for Aristotle's square of the circulating elements may have been astrological. After all, ancient circular cosmological diagrams of the stars and planets were calculating machines as well as maps. Further, their numerology evolved to a fourfold system based on the cardinal compass points (Festugiere 1944, vol. 1). They were intended to locate the movements of the heavenly bodies and to compute their significance for humanity. Adopting the Arabic astrological diagrammatic system, Ramon Lull in 1273–5 created a type of mechanically operated 'computer', the *Ars compendiosa*, also based on an elemental numerology of four. This was a series of concentric circles moderated by overlying squares or triangles at their centres. They were drawn on various sheets of paper which when moved about aimed to answer all questions, divine, moral, philosophical, logical and natural (Pring-Mill 1962: passim). Lull's diagrams also took the forms of highly schematic trees, ladders and precise tables of computations. All of these worked according to an alphabetical system distantly related to Arabic algebra and Hebraic cabbalism. It is easy to understand why this system offered a huge ray of hope to the alchemists and why they absorbed it into their own systems (Yates 1954: 115–73).

The alchemists would welcome anything which could 'map' their procedure for them, such as the Arabic-founded Lullism. Most of the time the alchemists were working completely blind, unsure of the actual nature of their substances or how to go about identifying them, lacking even thermometers which could tell them the temperatures of their fires, critical to the success of their operations. Since they had no means of quantification, they had to rely on the notation of qualities such as colour, smell and shape, as well as using astrological calculations of the movements of the heavens and the seasons (Multhauf 1966: passim).

Pseudo-Lullist alchemy adopted and considerably elaborated the original Lullist diagrams. But it also inherited from Lull an unresolved philosophical and religious conceptual dilemma. Lull had eventually developed two different numerological bases for the Lullian geometries, adding a numerology of three to the earlier four-fold one. These significantly altered his theory according to which one was applied. In alchemy, the two systems became alternative structures of the fundamental alchemical principles. They were either a symbolic numerology of 3 (the alchemical ontology of 'corpus–spiritus–anima') or of 4 (the four elements). It was difficult to reconcile them into one system and they functioned as parallel alchemical universes (Szulakowska 1988: 127–54).

Lull created a series of 'S'-circles (*Figura Animae S,* see Figure 4). Around the edges of the circles were placed the various 'powers of the soul' or the Dignities of God as established by neoplatonic Christianity. In their midst were overlaid four squares, representing the four material elements. God was allocated sixteen Dignities or qualities (4 × 4) at the edges of the

Figure 4 Florence, Biblioteca Nazionale-Centrale II iii 27, f. 229v(r), *Figura Animae 'S'* (Lullian) (*c*. 1470).

circles to accord with the elemental system of four governing the structure of the diagram (Szulakowska 1988: 127–30). In fact, God was given the structure of matter. Thus, to conform to the neoplatonic Augustinian theology of Catholic orthodoxy, Lull created another system in the *Ars Inventiva Veritatis* of 1289–90 based on an 'A' circle (the circle of the 'anima' or soul). The circle was mediated and computed by three triangles (3 × 3) and God was accordingly awarded nine Dignities. This then expressed the Catholic orthodoxy of God as being pure spirit in the three persons of the Catholic Trinity.

Pseudo-Lullian alchemy commenced with the *Testament* written in Catalan in Montpellier *c*. 1332. Ripley's major source is the *Testament* which, according to Pereira (1989: 22–5), was translated from Catalan into Latin probably by English alchemists in 1443. Thereafter, it continued to be the dominating influence on the English and focused their attention on the transmutation of metals.

The alchemists did not abandon the dubious, material pantheistic concepts of the 'S'-system, precisely because it expressed their own ideas and practice (Plessner 1975). The alchemists' theories include a strong dose of Stoic material pantheism in the concept of a single prime matter from which both body and soul arose (Barnes 1979: vol. 2, 38–44). Nonetheless, they also employed the triple numerology of the later *Figura Animae A* with its triangles to express their own alchemical trinity.

The pseudo-Lullists also employed diagrammatic tree or vine forms, from whose branches hung categories corresponding to those on related alchemical circles and which were given alphabetic equivalents. Trees became circles (Szulakowska 1988). In some manuscripts (such as British Library Ms Sloane 3667, f.52), Ripley adopted the tree or vine form and added his own alchemical terms to the various branches. These also exist as later sixteenth century engravings. However, Ripley's trees are straggly versions of the concise pseudo-Lullian versions. They are better described in their rambling disorder as 'maps' of the uncertain terrain of the alchemical process, or, perhaps, as laboratory notes, musings to himself as he worked. They relate to the castle of the *Duodecim Portae* only in general theory. Furthermore, whereas there exists this text by Ripley describing what is probably his version of a pseudo-Lullian 'S'-circle, there does not exist a drawing of one.

There are many extant pseudo-Lullian manuscripts of the *Testament* in the libraries of the United Kingdom which show images of the 'S'-circle or of other diagrams of the four elements (Singer 1928). The diagrams occur in relation to a section of the *Testament* known as the *Tractactatus de creatione mercuriorum ad faciendum tincturam rubeam* which Pereira (1989: 64–85) has established to be the basic text of the corpus (1989: 84). Another pseudo-Lullian text using diagrams and alphabets which was circulating England, certainly by the 1470s, was the 'Anima Artis' (*Compendium animae transmutationis metallorum*). (For example, Canterbury Cathedral Ms Lit. 8(50) is written in a Central Italian script of the 1470–80s and it may have arrived in the library at the same date. However, the '*Anima Artis*' is textually and visually more distant from Ripley's imagery.) Ripley could have seen these either in England or in Italy itself.

It would be an interesting exercise to construct Ripley's castle in the form of an 'S'-circle. Probably, there should be three quadrangles in the centre marked with the numbers of the gates from one to twelve at the corners, while the different alchemical stages from 'calcination' to 'projection' should be placed on the circumference of the circle. The points of the compass could then inscribe a square around the circle according to the pseudo-Lullian prototype. Following Ripley's instructions to always begin in the West, the quadrangles could be rotated to the different points of the compass and the changing correlates at the corners could be tabulated. This is a workable proposition but it is questionable as to what it would achieve. But then, this is not certain in the case of pseudo-Lullian diagrams either. Perhaps it is better to leave Ripley's castle as a general adaptation of the pseudo-Lullian format.

In addition, Ripley's castle could also relate, at least conceptually, to a missing diagrammatic tree which would be helpful in determining the exact route of Ripley's alchemical process at each gate. His alchemical journey is probably more complicated than the sequence of the poem suggests since he does not just enter each gate in turn but engages in subsidiary elemental voyages within them. Ripley gives certain clues when he describes the rotations. These go (three turns of the wheel each time) both clockwise and anticlockwise. In the second gate of Solution, Ripley moves west to north to east to south (clockwise). But in the tenth gate of Exaltation he travels west to south to east to north, i.e. against the sun.

But on what basis does Ripley arrive at twelve distinct stages of the alchemical process? Has he simply multiplied three by four and thus resolved the underlying ontological dilemma? Ripley states in the First Gate of Calcination the fact that each rotation must occur three times which would give him a multiple of three times the number of the four elements (Mangetus 1702: 276ff). The number twelve is not prominently featured in pseudo-Lullian alchemy. Lullian diagrams, whether 'A' or 'S', have related alphabets representing various categories of fifteen or sixteen letters (with variants) corresponding to a three or a four numerology respectively. Nor do Ripley's own vines have twelve categories. And in the Ripley *Scrowles* there are seven processes or stages (British Library Ms. Add, 5025, roll 4).

One possibility lies in Ripley's reference to 'altitude, latitude and profundity'. He could be thinking of a three-tiered geography: the plane of heaven, the plane of earth and the plane of hell. That there are heavenly elements is described by pseudo-Aristotle's *Metrologica* (Pagel and Winder 1974: 93ff) while the underworld, the gate of death, is part of the schema of alchemy.

There are two German treatises which could be the origins of Ripley's twelve processes: the *Aurora Consurgens* (Swiss, late 14C) has an image of the female sexual principle sitting in the midst of the signs of the Zodiac (Obrist 1982: 53); the *Rosarium philosophorum* (early 15C) shows the structure of the procedure by images of trees which bear human heads as 'fruit', of which there are twelve in the later stages (Fabricius 1976: 130, 162). Earlier treatises with such 'sacrificial trees' can have as many as fifteen or more 'heads' representing the processes. The highly influential and widely-dispersed *Rosarium* seems the most likely source of the number of Ripley's gates, though the *Rosarium* itself, like Ripley, was perhaps inspired to rationalise the alchemical procedure on the model of the circle of the twelve astrological signs. Ripley himself states briefly in the Fifth Gate that all things are engendered on earth through the rotation of the heavens (which is a statement found in the *Testament*).

A significant historical process of transformation and use of imagery is revealed. First, Ramon Lull turned Arabic astrological representations of the heavens into completely abstract logical diagrams. Then Ripley turned this pure abstraction back into a representation, first of a castle. Then from the idea of the movement of the computing pseudo-Lullian circles, Ripley began to evolve the concept of travel and of a navigational map of the earth. Working in Ripley's tradition and in that of his English successors Dee and Fludd, Maier completed the metaphor and turned it into a completely naturalistic terrestrial image: that of the circumnavigation of the globe. The important aspect of the above transmission of a mapping and voyaging theme is that the alchemists have turned Arabic astronomical and astrological maps into a map of the earth. Matter seems to have captured the heavens.

NOTES

1 J. J. Manget's edition of Ripley's *Liber Duodecim Portarum* is fundamentally Maier's translation from English into Latin published in his *Symbola Aureae Duodecim Nationum* (1617). Professor Stanton J. Linden of Washington State University is a pioneer in the study of Ripley, especially of the *Scrowles*. Hopefully, a publication of his findings will be made available in the near future.

2 Gate 1 = Calcination, 2 = Dissolution, 3 = Separation, 4 = Conjunction, 5 = Putrefaction, 6 = Congelation (Albefaction), 7 = Cibation, 8 = Sublimation, 9 = Fermentation, 10 = Exaltation, 11 = Multiplication, 12 = Projection. (Coudert 1980: 45–9 is a clear exposition of these processes.)

3 The rotations occur at Gates 1, 2, 3, 5, 6, 7, 9, 10.

REFERENCES

Ashmole, Elias (ed.) (1652). *Theatrum Chemicum Britannicum*. London.
Barnes, Jonathan (1979). *The Presocratic Philosophers*. 2 vols. London: Routledge and Kegan Paul.
Brann, N. L. (1979). George Ripley and the Abbot Trithemius. *Ambix* 26: 212–9.
Carbonelli, Giovanni (1925). *Sulle fonti storiche della chimia e dell'Alchimia in Italia*. Rome.
Clulee, Nicholas H. (1988). *John Dee's Natural Philosophy*. London and New York: Routledge and Kegan Paul.
Cornford, Francis M. (1980 [1912]). *From Religion to Philosophy*. Sussex: The Harvester Press.
Coudert, Allison (1980). *Alchemy*. Boulder: Shambhala.
Doberer, K. K. (1979). *The Goldmakers*. Connecticut: Greenwood Press.
Eliade, Mircea (1977). *Forgerons et Alchimistes*. Paris: Editions Flammarion.
Fabricius, Johannes (1976). *Alchemy*. Copenhagen: Rosenkilde and Bagger.
Festugière, André J. (1944). L'Astrologie et les sciences occultes. Vol. 1 of (1944–54): *La Révélation d'Hermès Trismégiste*. 4 vols. Paris.
French, Peter (1987 [1972]). *John Dee. The World of an Elizabethan Magus*. New York: Routledge and Kegan Paul.
Joachim, Harold H. (ed.) (1922). *Aristotle. On Coming-to-be and Passing-Away*. Oxford: Clarendon Press.
Jung, Carl G. (1980 [1953]). *Psychology and Alchemy*. London: Routledge.
Klossowski de Rola, Stanislas (1988). *The Golden Game: Alchemical Engravings of the 17th Century*. London: Thames and Hudson.
—— (1973). *The Secret Art of Alchemy*. London: Thames and Hudson.
Linden, Stanton J. (1990) *Image, Text and Meaning in the Ripley Scrolls*. Unpublished paper. Washington State University: Dept of English.
Maier, Michael (1618). *Atalanta fugiens*. Oppenheim: De Bry.
—— (1616). *De circulo physico quadrato*. Oppenheim: De Bry.
—— (1617). *Symbola aureae mensae duodecim nationum*. Frankfort-on-Main: Luca Jennis.
—— (1618). *Viatorium, hoc est, De montibus planetarum septem seu metallorum*. Oppenheim: De Bry.
Mangetus, Joannes Jacobus (ed.) (1702). *Biblioteca Chemica Curiosa* . . . 2 vols. Geneva.

Multhauf, Robert P. (1956). The significance of distillation. *Bulletin of the History of Medicine*. 439–46.
—— (1966). *The Origins of Chemistry*. London: Oldbourne.
Obrist, Barbara (1982). *Les Debuts de l'Imagerie Alchimiste* (XIVe–XVe *siècles*). Paris: Le Sycomore.
Pagel, Walter (1961). The prime matter of Paracelsus. *Ambix* 9: 117ff.
—— and Winder, Marianne (1974). The Higher Elements and Prime Matter in Renaissance Naturalism and in Paracelsus. *Ambix* 21: 93ff.
Pereira, Michela (1989). Warburg Institute Surveys and Texts XVIII. *The Alchemical Corpus Attributed to Raymond Lull*. London: The Warburg Institute.
Plass, Paul (1982). A Greek Alchemical Formula. *Ambix* 29: 59–72.
Plessner, Martin (1975). *Vorsokratische Philosophie und Griechische Alchemie*. Wiesbaden: Steiner.
Pring-Mill, R. D. F. (1962). *El microcosmos Lul. lia*. Oxford: Dolphin.
—— (1963). *El Numero Primitivo de Las Dignidades en el 'Arte General'*. Oxford: Dolphin.
Purce, Jill (1974). *The Mystic Spiral*. London: Thames and Hudson.
Ripley, George (1591). *The Compound of Alchymy*. London.
—— (1470). London, British Library Ms Sloane 3667, *Opera Chemica*.
—— (1649). *Opera Omnia Chemica*. Kassel.
—— (1588). London, British Library Ms Additional 5025, *Scrowle*. 4 rolls. 'Lubeck'.
Roberts, R. J. and Watson, A. G. (1990). *John Dee's Library Catalogue*. London.
Sheppard, H. J. (1957–8). Gnosticism and Alchemy. *Ambix* 6: 140–48.
—— (1960). The Redemption Theme and Hellenistic Alchemy. *Ambix* 8: 98–110.
—— (1962). The Ouroboros and the Unity of Matter in Alchemy. *Ambix* 10: 83ff.
Sherwood Taylor, F. George Ripley's Song. *Ambix* 2: 177–82.
—— (1976 [1952]). *The Alchemists*. St Albans: Paladin.
Silberer, Herbert (1971 [1917]). *Hidden Symbolism of Alchemy and the Occult Arts*. New York: Dover.
Singer, Dorothea Waley (1928). *Catalogue of Latin and Vernacular Alchemical Manuscripts in Great Britain and Ireland dating from before the sixteeth century*, vol. 1, Brussels.
Stapleton, H. E., Lewis, G. L. and Sherwood-Taylor, F.

(1949). The Sayings of Hermes quoted in the MA AL-WARAQT of Ibn Umail. *Ambix* 3: 69–90.

Szulakowska, Urszula (1986). The Tree of Aristotle: Images of the Philosophers' Stone and their Transference in Alchemy from the Fifteenth to the Twentieth Century. *Ambix* 33: 53–77.

—— (1988). Thirteenth Century Material Pantheism: in the Pseudo-Lullian 'S'-Circle of the Powers of the Soul. *Ambix* 35: 127–154.

Theatrum Chemicum. (anon ed.) (1602). Vols 1–3. Ursel.

Thorndike, Lynn (1953). *A History of Magic and Experimental Science* Multivol. Second Reprint. New York: Columbia UP.

Widengren, Geo (1965). *Mani and Manicheism*. London: Weidenfeld and Nicholson.

Yates, Frances A. (1954). The Art of Ramon Lull. *The Journal of the Warburg and Courtauld Institutes*, 115–73.

—— (1960). Ramon Lull and John Scotus Erigena. JWCI, pp. 1–44.

—— (1986 [1972]). *The Rosicrucian Enlightenment*. New York: Routledge.

CYRIL WILLIAMS

Waldo Williams: A Celtic Mystic?[1]

Opinions differ as to whether all forms of mysticism are the same in essence, notwithstanding differences in terminology and cultural background. For example, do terms like satori, nirvana, śūnyatā, the kingdom of heaven within, signify an identical experience? If that is so, can one legitimately speak of types of mysticism or use designations such as Celtic or Welsh mysticism? The term mysticism itself has been variously applied. Like the word 'mystery' it is derived from the Greek root *muéin* (Principe 1976: 2). In Christian circles, it came to mean direct experiential knowledge of God, but such a description would need qualification, of course, in referring to mystical experience in non-theistic contexts, such as Theravāda Buddhism or indeed Vedāntic non-dualism or philosophical Taoism. In all three, there is a pronounced sense of reality which is not understood in personal theistic terms. Are we then to say of mysticism as the Rig-Veda said of Reality 'Truth, i.e. Reality, is One but wise men give it different names'? In which case, Celtic mysticism, Chinese mysticism and Indian mysticism would all signify the same identical experience. Some similarities of descriptions and common elements from within the various traditions would seem to support such a view.

In considering any experience, however, three aspects are to be borne in mind. There is the experience itself, the interpretation and understanding of the experience by the subject and the interpretation given by the third party, who may be better informed or more qualified by virtue of wider familiarity with accounts of spiritual experiences than the subject himself or herself. Even so, the experience of the ultimate itself eludes the most sophisticated research and map making and there is also the fact that one can never enter fully into the experience of another. But claims have been made that through heightened perception or pure consciousness a pure experience of the ultimate is attainable. It is held to be the meeting point where all differences disappear, a nothingness beyond all discrimination, an experience of what is beyond experience, a 'wholly other' empty of all categories. If such a state can be, or has been reached, I cannot say. John Hick regards the differences as culturally relative signs of a higher reality which impinges on man's consciousness (Coward and Penhelum 1976: 71). Case descriptions of mystical experiences generally refer to experiences on lower rungs of the ascending ladder, where the ladder itself is still necessary equipment in the ascent for those who remain in the realm of limited attainment. It is the cultural background which provides the ladder, or, to make a bolder claim, the cultural conditioning is inseparable from the experience, shaping it, permeating it.

S. T. Katz does not hesitate to go further than this to deny the occurrence of *pure* (i.e. unmediated) experiences. He writes: '. . . in order to understand mysticism it is *not* just a question of studying the reports of the mystic after the experiential event but of acknowledging that the experience itself as well as the form in which it is reported is shaped by concepts which the mystic brings to, and which shape his experience' (Katz 1978: 26). Again, '. . . the forms of consciousness which the mystic brings to experience set structured and limiting parameters on what the experience will be'.

Accepting this view, one could speak of a Celtic mysticism and within it of the mysticism of component units including Welsh mysticism. At the same time, one is very aware that there are affinities and resemblances which cross cultural divides even at the descriptive level and of yearnings of the spirit which unite kindred souls of diverse backgrounds. Yet within the family of mystics at the descriptive levels, at any rate, one can discern distinctive cultural moulds and colour. Whether there is an experience which transcends the experiencer only the mystic who has arrived can say, and even to say it means to descend the ladder which has been kicked away! Thus an impasse is reached where only silence or the 'neti, neti' of the Tathagata is appropriate in relation to the ultimate, unless the ultimate deigns to communicate with us in our respective spheres where descriptions have significance. This, of course, is the claim of prophetic religions. What is revealed of Reality, although only partially understood, is deemed to be of the same nature as the ultimate as it is in its wholeness. For instance, to say that the nature of the Absolute or the Ultimate is Love is to claim that it is discernible, however dimly and partially understood, experientially. The part that is 'known' is identified by the experiencer with the Ultimate and communicated in culturally determined forms. It would be misleading, however, to discriminate starkly between prophetic and mystical claiming that in the former the revelatory experience is 'given' while, in the latter, experience of reality is something which is striven for. In the prophetic stance, too, there can be a conditioning of receptivity and in the mystical way a sense of a Reality 'take-over' which is in no way the product of preparation.

In the case of Waldo Williams, his understanding of the ultimate is unseverable from his deep sense of the unity of humanity. One also finds in his poems, on occasion, a sense of the unity of the whole cosmos – a feature, according to some, of the Celtic tradition but which is certainly not absent elsewhere, particularly in Indian literature, such as the Upanishads. I do not propose here to consider the question of Celtic heritage in general, or even the more basic question of whether the term Celtic is merely a matter of convenience to indicate a mixed population in several countries. I dare say new methods of study in genetics, carbon dating and linguistic studies may lead to the revision of some earlier theories. Again, even within more closely related groups, such as the P or Q Celts, generalisations are fraught with hazards. Yet certain traits are discernible in the legendary material which has come down to us in the nations designated Celtic – enough, perhaps, to speak of characteristics. Some writers, for instance, have remarked upon the sense of mystery, the consciousness of unity and the interconnectedness of the whole cosmos, belief in the continuation of life through different forms and in a dialogue and relationship between two worlds signifying one all embracing world. Often mentioned also is the melancholia of the Celtic spirit, deriving from ever striving for a goal which is beyond reach; the sense of the incompleteness of each and everything and the Celt's way with words with its concern for form as well as content, especially the sound of words. The difficulty, of course, is that we are in the realm of speculation depending not on written records for the early story but on archaeological finds and anthropological theories. In fact, Professor Kenneth Jackson dismissed all talk about melancholia and the propensity of the Celt to mysticism as nonsense. He wrote on the basis of the evidence of later literature: 'In fact, the Celtic literatures are about as little given to mysticism or sentimentality as it is possible to be; their most outstanding character is rather their astonishing power of imagination' (Jackson 1971: 20).

On the other hand, Dr Pennar Davies, in *Rhwng Chwedl a Chredo*, a title which might be rendered 'Fable and Faith', speaks of recurring motifs in the ancient Brythonic culture as in the wider

Celtic world, which he claims are not incompatible with basic attitudes discernible in Celtic Christianity, earlier Welsh poetry, medieval literature, and in that radicalism which he regards as an unmistakable part of the Welsh heritage. At the same time, he is aware of the temptation to read a pan-Celtic significance into each instance of seemingly parallel forms. This is a salutary word of wisdom from one who is well aware of the intermingling of cultures and influences.

Turning to the Welsh tradition, some have found a strong vein of Neo-Platonism, tracing influences from Plotinus to Dionysius in the early centuries, then through people like Bernard of Clairvaux, Bonaventura, Meister Eckhart and Jacob Boehme in the sixteenth century, and hence to Morgan Llwyd who helped to found the first Independent church in Wales at Llanfaches in 1639. This is bold map-making indeed, but if one wanted to be even bolder one could speculate on influences on Plotinus himself. He had been a student at Alexandria, the crossroads of Eastern and Western cultures, and we also hear of a journey he made to Persia expressly to familiarise himself with the teachings of the East. Persia had imbibed from the same or similar font of religious thought as the Indian traditions, as a comparison of the Avesta and the Vedas show, and of which there are traces also in Celtic mythology as well as, I believe, linguistic links. I mention this wider canvas and global diffusion, for Waldo Williams, immersed in Celtic lore with its Arthurian legends and enamoured of the magic of his beloved Pembrokeshire though he was, yet had imbibed also from the fonts of classical, European and Indian literatures. Roots and branches, both his Welsh heritage and his deep sense of the unity of the human race, are prominent in the catholicity of his thinking. Yet objection has been expressed to journeying 'so far afield as the Eleusinian mysteries and Indian tree myths' to explicate his symbolism (Bianchi 1978: 8).

He is one of the leading Welsh poets of our time. Deeply rooted in the soil of cairns, stone circles and cromlechs, yet his square mile is by no means parochial but extends world-wide. Henry Vaughan, Gerard Manley Hopkins, Coleridge, Keats, Russian novels, the Upanishads, the Gītā, all are familiar territories for him. Although he has left us but relatively few poems, in recent years more has been written about him than any other poet on the Welsh scene. In my opinion he can be described as a mystic but the description needs elaboration since he confutes some popular erroneous impressions of mystics, combining his inward spirituality and sensitivity to the unseen with a keen sense of humour and a deep concern with mundane problems, social injustice and efforts to establish world peace, proving yet again that many so-called other-worldly mystics are practically orientated and persons committed to social action.

More often than not, in the terminology of philosophical theology, the description theistic mysticism could apply to him, yet on occasion his awareness of the unity of all would make monistic mysticism more appropriate. But if the flight of the mystic takes one beyond good and evil, then Waldo was no mystic, for his spirituality is one which can never be divorced from its grounding in morality, and it has to find expression in moral activity. His poems are related to life and the noble world of ordinary people motivated by a strong sense of social justice. This no doubt reflects in part the influence of his father, whose heroes included William Morris, Walt Whitman, Keir Hardie, Shelley and Tolstoy (Thomas 1985: 8). But Waldo, while ultra-sensitive to international injustices and strife, like many contemporary Welsh poets, eventually found his own political home in Plaid Cymru and on one occasion reluctantly offered himself as a parliamentary candidate, the most unlikely aspirant in any election!

As a Quaker, perhaps it is not surprising that he finds the transcendent within. But the inner experience demanded of him obedience in action. He was stung to the core by Gandhi's admonition of Tagore, 'You give us deeds not words' (Rhys 1981: 11). During the Korean War, Waldo was sorely troubled by the inhumanity of mankind and by the thought that the contributions of ordinary people to the state, including his own, constituted a material factor in the pursuit of the war and the killing entailed. In consequence, translating conviction into action, he withheld from Inland Revenue the portion of his tax which he calculated was used for destructive purposes and gave up his teaching post for the same reason. Inevitably, he was brought to court and eventually sent to prison on two occasions, once in 1960 and again the following year.

The whole experience was a kind of catharsis on his part and until this had happened the poet had been silent for a period of some three years. When he broke the silence he wrote the now famous poem, and one of the most intriguing, perhaps, in the Welsh language. It is entitled 'Mewn Dau Gae' which Joseph Clancy and Anthony Conran have rendered 'In Two Fields'.[2] I shall return presently to this poem not only because I find it fascinating, but also because, in reading again Deirdre Green's contribution, 'Living Between the Worlds', in Karl Werner (ed.) *The Yogi and the Mystic* (1989: 121–39) and her volume on St Teresa of Avila (1989), I believe, with her increasing interest in and knowledge of the background of Welsh literature, it would greatly appeal to her, and she would recognise the mystic in the paradoxes of the poet and his experience.

Lest I be accused of prejudice in underlining the attraction and power of sound and form in Welsh poetry, I shall let Clancy speak for me! He has shown remarkable industry and dedication in translating Early, Medieval and Modern Welsh poetry, but is only too willing to recognise that the melodic patterns of the original cannot be reproduced in translation. As an illustration of some of the problems, he notes the very last line of 'In Two Fields', a poem in free verse. It reads:

> Daw'r Brenin Alltud a'r brwyn yn hollti

He comments: 'the powerful image of the coming of the Exile King and the rushes falling apart before him is intensified by the sound pattern, known as *cynghanedd groes o gyswllt*, i.e. the consonants in the first half of the line – d, r, br, n, n, ll, t – begin to be precisely matched just before the mid-line pause, so that the "d" in "Alltud" is followed by r, br, n, n, ll and t. Even if one could duplicate this type of consonantal patterning in English, it would be impossible to do so while approximating the meaning of the line' (Clancy 1982: xli).

To evade the *cynghanedd*, in Waldo's own view, would be to produce something quite different to the original. True, we have what he called consonantal chiming in Manley Hopkins, but *cynghanedd* is a much richer technique than alliteration. It imposes order and introduces a deeper relation between the parts of a poem's line when form and meaning coalesce and embrace each other or, in the words of the old bardic schools, the 'sweetness to the heart' derived from the 'sweetness to the ear' (Waldo 1953: 16). In the perfect fusion of literary form and content, approaching 'the condition of music', the meaning is beyond translation. In fact, translating from any language to another in most cases is to engage in the art of the impossible.

In August 1952, Waldo Williams addressed the Conference of Extra-Mural Tutors at Aberystwyth on 'The Function of Literature' (Waldo 1953: 12–18) and, although the topic was more general than poetry, many of his observations provide insights into his thinking on the nature and function of poetry. He refers to its rhythm as the distinctive characteristic of poetry: 'It calms the tumbling emotion and objectifies it.' Sound is all-important for Waldo:

> The effect of sound can reach through the physical beings more powerfully than that of sight. The drum can lead us when the flag can not. Add to this the strange duality of

language – outwardly a net thrown over the world and yet such an intimate faculty as to be part almost of the mind itself. Little wonder it is that a mystique of language has sometimes arisen.

For Waldo, the poetic image and the feeling, when they come, come together, and find expression in a speech continuum which is voiced inwardly as it always must be. The bulk of his poetry, according to J. Caerwyn Williams, is a dialogue between the poet and himself. We are privileged to listen in on this dialogue, but are we free to introduce our own interpretation irrespective of whether it differs from the poet's intention? It is a commonly held view that once a poem or a play is launched then, to continue the nautical metaphor, it may bear whatever cargo the auditor wishes it to convey. It may bear meanings beyond the author's ken.

In a letter to *Baner ac Amserau Cymru* (13: ii: 1958) the poet gives his own explanation of the poem, 'In Between Two Fields'. This was in response to the complaint concerning its obscurity. Had he thought it was obscure, he said, he would not have written it, adding:

> There is abroad today a theory that several interpretations of a poem are possible and that those of which the author was unaware are as good as, or even better sometimes, than his own. I would find no pleasure in composing in this spirit. The difficulty with a theme like this, which constitutes large chunks of experience, is that experience would be weakened and distorted by a systematic expression of the ideas, losing the excitement. They did not come to me like that. I sought to do one thing: leave plenty of switches here and there on the walls believing that the whole room would be illuminated if the groper would touch one of them.

This means, says Caerwyn Williams, that effective communication is essential in poetry. Moreover this depends on the integrity of the poet's effort to commune with himself, for it is only thus he can commune with his readers. True, but the poet or dramatist or hymn writer may still speak more than he or she knows. A poem comes tied in with a specific culture and the experience which brought it about may be documented, but inspired thought, though coloured by the culture that occasions it and providing us with elements of recognition, yet also, in part at least, transcends the culture that brought it into being and may come 'alive' to an auditor in another context. A poem may speak in different vein to different listeners. This becomes increasingly evident in the case of Waldo, in spite of his protestation, as the amount of critical appraisal grows. Now almost a cult figure, he has been appropriated by diametrically opposed theological camps.

What then of 'In Two Fields'? In his explanation, Waldo names the farm, the owner, and the precise location where he had that vivid experience forty years previously, which would make him about fourteen at the time. He was in the gap between two fields belonging to John Beynon, The Cross, Clunderwen, Pembrokeshire, when in his own words, 'I suddenly realised, and in a very real sense, (*and*) on a pronounced personal occasion, that men are all brothers of one another.' He goes on to describe a state of joy when all around becomes alive. In the poem, he said it was like wells springing up into the air as fountains and the drops falling back as leaves to earth. But the printer had rendered 'dafnau' (drops) as 'dagrau' (tears), so the poem read:

> There were fountains bursting towards the heavens
> And falling back, their tears like the leaves of a tree.

Waldo did not change it and even considered it an improvement, depicting the tears as the release of feeling.

His mysticism which interiorises the transcendent also externalises it in the strong sense of the corporate community. Brotherhood was indivisible from Reality for Waldo. This is attested in his statement to the Objectors Tribunal at Carmarthen, February 12th 1942:

> I believe all men to be brothers and to be humble partakers of the Divine Imagination that brought forth the world, and that now enables us to be born again into its own richness by doing unto others as we would others do unto us . . . .
>
> I believe Divine Sympathy to be the full realisation of the Imagination that brought forth the world. I believe that all men possess it obscurely and in part, and that it has attained its perfect expression in the life and teaching of Jesus . . . .
>
> I believe that the Spirit communes not with societies as such, but directly and singly with the souls of men and women, thereby enabling us to commune fully with each other, forming societies. I believe, therefore that my first duty to the community to which I belong is to maintain the integrity of my own personality. 'God is a spirit and they that worship Him must worship Him in spirit and in truth'.
> (Nicholas 1975: 121; 1975: 44–5)

He strikes a similar note in an important article entitled 'Brenhiniaeth a Brawdoliaeth' (Kingship and Kinship) which illustrates his familiarity with the work of Nicholas Berdyaev and Tolstoy. The harmony of true brotherhood suggests for him the

Will of another, a Third. This, perhaps, is where he is more explicit than Buber, for while it would seem that we find echoes of *Ich und Du* in Waldo's thought, and while the Third is not absent in Buber, the Third is more explicit in the poet. This is brought out in his poem entitled 'Brawdoliaeth' (Fraternity) written some time during the early part of the war, 1940(?):

> God's hidden net
> Holds every living person;
> Reconciliation and a full web
> I, Thou, He.

There is another poem, without title, composed in bereavement after the loss of his wife, Linda. It is, in spite of, or indeed because of, his grief, a song of hope which speaks not only of a basic unity but also of the restorative power of Life at the heart of Beingness itself:

> There is no withering in the root of Being
> There our pith is retained,
> There is the courage that is gentleness
> The life of every frail life.
>
> It is there that the heart retreats after the storm,
> The world is shattered.
> But in the low fortress the squirrel of bliss
> Tonight makes its nest.
>   (Nicholas 1975: 59)

Brotherhood without; a sense of a Third presence, confidence in the depth of Being, Waldo has also a profound feeling for the relatedness of all things. The root symbol is always prominent, and so are branches, which reminds us, as Caerwyn Williams says, of the prominence of the cosmic tree in earliest literatures and particularly in Indian literature. Thus Waldo asks:

> What is knowing? Having one root
>   Under the branches.[3]

While Waldo seems to depart from his theistic framework on occasion, generally it is the personal ultimate which is supreme, just as Krishna, with disregard for the orderliness of human logic, is supreme above the Absolute Brahman in sections of the *Gītā*. Even 'Awen', the Inspirer, is personalised in the poem *Adnabod* and is addressed almost as God. It is a Thou which is the means, the end, the art of knowing, and cannot be reduced to the upsurges of the unconscious:

> Thou art our breath. Thou art soaring
> Of our longing to the deep sky.

Thou art the water which runs
From the barrenness of anxiety and fear.
Thou art the salt to purify us.
Thou art the wind that cuts the pomp about us.
Thou art the wayfarer that knocks.
Thou art the prince that abides in us.

In his dialogue with himself, he asks the question of the ages, a question asked by the ancient sages of the Vedas, by Gautama the Buddha, by philosophers in many lands: *Who am I?* Waldo gives an answer which is Zen-like in nature:

Who am I? Who am I?
A stretch of my arms and there, between their two stumps
The dread of thinking about myself
And questioning the ground of all questioning
Who is this?
The sound of water, I wade it for an answer.
Nothing but the cold current.
Through the ditch homewards if there is a homewards,
I heft the gate-post still doubting,
And O, before I reach the back-door,
The sound of building a new earth, new heaven,
Were my mother's clogs on the kitchen floor.

The lines are from the poem 'A Summer Cloud' where the poet is overwhelmed by the enormity of a world too silent for living:

No there. Only me is here
Me
With no father or mother or sisters or brother,
And the beginning and end closing about me
                              (Clancy 1982: 130f.)

Then he is rescued by the sight of the familiar and the answer to his need.

Time is also a frequent thought in Waldo's poems, and he has one entitled 'The Moment', where he is conscious of timeless time which is not mundane time yet comes to us in that, and there is no mistaking the mystical quality of the experience which inspires it:

The Moment

The Moment is a topic
No scholar writes about.
The river halts its flowing
And the rock cries out
To give its sworn word

To things eye has not seen
And ear has not heard.

Breeze among the breezes,
Sun beyond the sun,
The true homeland's wonder
Unwarped, unworn,
Seizing earth in its power –
We know by the Moment's coming
We are born for the hour.[4]

I trust that the examples given have conveyed something of what I regard as the mystical quality of the poems, but as already stated, the best translation is bound to fail. As the poet himself has said of his land and language:

Here are the mountains. One language alone can lift them
And set them in their freedom against a sky of song.

As in the sacred mantra where the silence at the end of the *AUM* is powerfully evoked by the mantra itself, so does the quality of sound in the poet's words evoke what cannot be conveyed adequately. We cannot enter the experience of the young poet 'In Two Fields', but reading it in his own tongue can become an experience also.

NOTES

1 A paper read on 4 September, 1992 at a conference of the Traditional Cosmology Society held at Saint David's University College, Lampeter, to honour the memory of Dr Deirdre Green, lecturer at the College from 1985 to 1990.
2 The poem is in *Dail Pren* (Gomer 1956), and Clancy's translation in *Twentieth Century Welsh Poems* (Gomer 1982), p. 135.
3 'Beth yw adnabod? Cael un gwraidd
    Dan y canghennau!'

   'What is recognition? To find one root
    Beneath the branches!'
                                    (trs. Nicholas 1975: 60)

   'Abnabod' is 'knowing of', or 'knowing a person in a relationship', while the Welsh word for knowing about or knowing a fact is 'gwybod'.
4 Clancy 1982: 136. See also Anthony Conran's translation in *The Penguin Book of Welsh Verse*.

REFERENCES

Bianchi, Tony (1978). Waldo and Apocalypse. *Planet* 44. Llandysol: Gomer.

Clancy, J. P. (ed.) (1982). *Twentieth Century Welsh Poems*. Llandysol: Gomer.

Conran, A. (ed.) (1967). *The Penguin Book of Welsh Verse*. London: Penguin.

Coward, H. and Penelhum, T. (eds) (1976). *Mystics and Scholars: The Calgary Conference on Mysticism*. Waterloo, Ontario: Wilfrid Laurier University Press.

Green, Deirdre (1989). *Gold in the Crucible: Theresa of Avila and the Western Mystical Tradition*. Shaftsbury: Element.

Jackson, K. H. (1971). *A Celtic Miscellany*. London: Penguin.

Katz, S. (ed.) (1978). *Mysticism and Philosophical Analysis*. New York: Oxford University Press.

Nicholas, J. (1975). *Waldo*. Writers of Wales Series. Cardiff: University of Wales.

Principe, Walter H. (1976). 'Mysticism: its Meaning and Varieties'. In Coward and Penelhum (eds) (1976). *Mystics and Scholars: The Calgary Conference on Mysticism*. Waterloo, Ontario: Wilfrid Laurier University Press.

Rhys, Robert (ed.) (1981) *Waldo Williams*. Swansea: Christopher Davies.

Thomas, Ned (1985). *Waldo*. Caernarfon: Pantycelyn.

Werner, Karel (ed.) (1989). *The Yogi and the Mystic*. London: Curzon Press.

Williams, Waldo (1953). *Yr Einion: The Welsh Anvil*. University of Wales Guild of Graduates.

—— (1956). *Seren Gomer*. Llandysul: Gomer.

—— (1971). *Y Traethodydd*. Caernarfon: Llyfrfa'r Methodistiaid Calfinaidd.

ROY WOODS

Against Mapping Invisible Worlds in Rilke's *Duino Elegies*

Whilst trying recently to explore Yorkshire in the winter fog, I appreciated once more how useful maps are, and yet also realised more fully how they can distract us from experiencing the real world. This also seemed a suitable metaphor for spiritual search. How many of us read spiritual guide books rather than making the real journey? Books on mysticism or Zen, for example, can seem most rewarding but do they actually give even the most fleeting glimpse of the reality they point to? Rather than providing help on the journey, they can become substitutes for it. Without a map we risk getting lost – and not only in the fog – but staying at home studying maps, or devoting too much attention to them on our actual journey can prevent us from experiencing and appreciating the landscape we are part of.

The poet Rainer Maria Rilke (1875–1926) came to appreciate this problem more and more as he attempted to map the invisible by translating his experience into poetry. He did not, as far as I know, use the concept of mapping but he did feel strongly that our *interpretations* are an evasion and disinherit us from our true spiritual home. In his major work, the *Duino Elegies*, he conveys our alienation by contrasting us with a series of figures who are not, in his view, alienated; and this includes animals. In the first Elegy he put it like this:

>[. . .] the noticing beasts are aware
>that we don't feel very securely at home
>in this interpreted world.
>   (Leishman and Spender 1963: 25)

>[. . .] die findigen Tiere merken es schon,
>daß wir nicht sehr verläßlich zu Haus sind
>in der gedeuteten Welt.
>   (Rilke, *S. W. I* 1955: 685)

Paradoxically, of course, Rilke's reading of the situation in the Elegies is itself yet another level of interpretation. Nonetheless he persevered with his diagnosis of our alienation: interpretations, our preconceived notions, our 'maps', he saw as interfering with true perception. They place us 'outside' the world, 'over and against' it, we remain 'opposite everything'. By thought and by 'normal' perception we divide ourselves from the unity of which we are really a part. Rilke did not *reason* his way to this perception; he saw it directly. For him – just as for that great pragmatist Gautama, the Buddha – seeing was far more important than thinking. Experiential, not intellectual, discovery was what counted.

Rilke had several 'mystical' experiences which convinced him of the error in our usual divisive way of perceiving. One such 'encounter' took place in 1912 in Duino Castle high on the cliffs overlooking the Adriatic and this provided the initial impetus for the *Duino Elegies (Duineser Elegien* in Rilke 1955 and Leishman and Spender 1963). It is said that he heard the first lines of several of the elegies dictated to him on the wind from the unitive dimension he called 'the angelic orders' (*der Engel Ordnungen*). In a letter to Lou Andres Salomé he objectivised the experience, distancing himself from it by using third person description:

> He remembered the hour in that other southern garden (Capri) when, both outside and within him, the cry of a bird was correspondingly present, did not, so to speak, break upon the barriers of his body, but gathered inner and outer together into one uninterrupted space, in which, mysteriously protected, only one single spot of purest, deepest consciousness remained. (L. & S. 1963: 155)

This perception of the mystical unity of inner and outer was to become a central feature of his later poetry, but it also proves to be the main source of the notorious 'difficulty' of his work. His monistic interpretation of reality clashes with the dualistic nature of our language and thought patterns. We habitually dissect the world into discrete entities.

Although he was a poet and depended on language and thought as essential tools, Rilke, like the Buddhists in Asia and the Taoists in China, considered these to be a hindrance to true understanding and direct experience of the world. Because of our habitual thought and language patterns, we separate ourselves from our environment and divide everything into mutually exclusive opposites, thus failing to appreciate their inter-dependence.

We miss the fact that light, for example, depends on darkness, good on evil, male on female, existence on non-existence, and so on.

The most stubborn of these polarities remains the Cartesian duality of subject and object. We assume that the model of a perceiving 'subject' looking 'out' on a perceived 'object' is the way things really are, particularly as the syntax of most Western languages reinforces this dualism, but experience proved to Rilke that this was merely a convention standing in the way of 'unmediated vision'.

What makes his work even more problematic and, for some, difficult to accept, is the way he took his rejection of duality one stage further: he also eliminated the distinction between life and death. Using the image of the unilluminated side of the moon – invisible but real – he conveyed death as unseen but real, a part of life, not apart from it. With purer perception these two poles, Rilke maintained, become parts of the same unified whole. He tried to explain this in a letter to his Polish translator, Witold von Hulewicz:

> ... in the (Duino) 'Elegies' affirmation of life AND affirmation of death reveals itself as one. To concede to the one without the other is [. . .] a restriction that finally excludes all of infinity. Death is that side of life which is turned away from us, unilluminated by us: we must try to achieve the greatest possible consciousness of our existence, which is at home in both of these unlimited provinces, inexhaustibly nourished out of both [. . .] the true form of life extends through both regions, the blood of the mightiest circulation pulses through both: *there is neither a here nor a beyond, but only the great unity*, in which the 'Angels,' those beings that surpass us, are at home.   (L. & S. 1963: 110)

In a poem written in 1914 he coined the term '*Weltinnenraum*' (world-inner-space) to describe this realm of pure consciousness:

> There reaches through all beings one single space:
> world-inner-space. Silently birds fly
> right through us. Oh, I who wish to grow,
> I look out and within me grows the tree.
>   (my translation)

> Durch alle Wesen reicht der *eine* Raum:
> Weltinnenraum. Die Vögel fliegen still
> durch uns hindurch. O, der ich wachsen will,
> ich seh hinaus, und *in* mir wächst der Baum.
>   (*S. W. II* 1955: 93)

He expresses a similarly heightened, unified consciousness in another poem written later, 1924. It is interesting that here, as in Buddhist *śūnyatā* (the void) and in subatomic physics, *space* is conceived as an active force rather than a mere passive container:

> Space reaches out of us and translates things:
> that the essence of a tree succeed you
> hurl inner space around it from that space
> that's present within you.
>
> (my translation)
>
> Raum greift aus uns und übersetzt die Dinge:
> daß dir das Dasein eines Baums gelinge,
> wirf Innenraum um ihn, aus jenem Raum,
> der in dir west.
>
> (S. W. II 1955: 168)

In the Elegies Rilke inverts the traditional order of things given in *Genesis*, so that man, because of his distorted perception, is no longer seen as superior to the rest of creation, he is placed at the bottom of the ladder of perfection with the angels at the top and a range of figures in between, all of whom overshadow him, for they are more at home in 'world-inner-space' or 'the Open', as he also called it.

Much has been written about these figures and I have no wish here to cover old ground, but just to clarify Rilke's hierarchy briefly, they are: the young dead, lovers, dolls, the hero, animals and, finally, things. None of these has man's problem of standing outside the world and dissecting it into opposites. As already shown: instead of simply *being*, as animals and roses do, we *interpret* and thus fail really to perceive. We are never here in the present, we are always separated, 'always opposite'.

In the first of the ten *Duino Elegies* Rilke depicts how alien we would find it if we did throw off our dualistic interpretations and experience the Open (although he conveys it in the indicative rather than the conditional):

> True, it is strange to inhabit the earth no longer,
> to use no longer customs scarcely acquired,
> not to interpret roses, and other things
> that promised so much in terms of a human future;
> to be no longer all that one used to be
> in endlessly anxious hands, and to lay aside
> even one's proper name like a broken toy.
> Strange, not to go on wishing one's wishes. Strange,
> to see all that was once relation so loosely fluttering
> hither and thither in space. And it's hard, being dead,

and full of retrieving before one begins to perceive
a little eternity. – All of the living, though,
make the mistake of drawing too sharp distinctions.
Angels (it's said) would be often unable to tell
whether they moved among living or dead.
(L. & S. 1963: 29)

Freilich ist es seltsam, die Erde nicht mehr zu bewohnen,
kaum erlernte Gebräuche nicht mehr zu üben,
Rosen, und andern eigens versprechenden Dingen
nicht die Bedeutung menschlicher Zukunft zu geben;
das, was man war in unendlich ängstlichen Händen,
nicht mehr zu sein, und selbst den eigenen Namen
wegzulassen wie ein zerbrochenes Spielzeug.
Seltsam, die Wünsche nicht weiterzuwünschen. Seltsam,
alles, was sich bezog, so lose im Raume
flattern zu sehen. Und Totsein ist mühsam
und voller Nachholn, daß man allmählich ein wenig
Ewigkeit spürt. Aber Lebendige machen
alle den Fehler, daß sie zu stark unterscheiden.
Engel (sagt man) wüßten oft nicht, ob sie unter
Lebenden gehn oder Toten.
(*S. W. I* 1955: 687f)

We are taught the subject–object dichotomy in early childhood. That marvellous intensity which most of us have already forgotten, is soon eliminated. As Rilke puts it in the Eighth Elegy:

for while a child's quite small we take it
and turn it round and force it to look backwards
at conformation, not that openness
so deep within the brute's face. Free from death.
We alone see *that*; the free animal
has its decease perpetually behind it
and God in front, and when it moves, it moves
within eternity, like running springs.
We've never, no, not for a single day,
pure space before us, such as that which flowers
endlessly open into.
(L. & S. 1963: 77)

Denn schon das frühe Kind
wenden wir um und zwingens, daß es rückwärts
Gestaltung sehe, nicht das Offene, das
im Tiergesicht so tief ist. Frei von Tod.
Ihn sehen wir allein; das freie Tier

> hat seinen Untergang stets hinter sich
> und vor sich Gott, und wenn es geht, so gehts
> in Ewigkeit, so wie die Brunnen gehen.
> *Wir* haben nie, nicht einen einzigen Tag,
> den reinen Raum vor uns, in den die Blumen
> unendlich aufgehn.
>
> <div align="right">(<i>S. W. I</i> 1955: 714)</div>

He further elucidates the 'Open' in a description of the Spanish landscape, written in 1915: 'Everywhere appearance and vision came, as it were, together in the object, in every one of them a whole inner world was exhibited, as though an angel, in whom space was included, were blind and looking into himself. This world, regarded no longer from the human point of view, but as it is within the angel, is perhaps my real task.' (L. & S. 1963: 10). And in another letter in 1925 he wrote:

> You must understand the concept of the 'Open', which I have tried to propose in the [eighth] elegy, in such a way that the animal's degree of consciousness sets it into the world without the animal's placing the world over against itself at every moment (as we do); the animal is *in* the world; we stand *before it* by virtue of that peculiar turn and intensification which our consciousness has taken. [. . .] By the 'Open', therefore, I do not mean sky, air, and space; *they*, too, are 'object' and thus 'opaque' and closed to the man who observes and judges. The animal, the flower, presumably *is* all that, without accounting to itself, and therefore has before itself and above itself that indescribably open freedom which perhaps has its (extremely fleeting) equivalents among us only in those first moments of love when one human being sees his own vastness in another, his beloved, and in man's elevation towards God.   (Heidegger 1975: 98)

All the purer beings of Rilke's *Duino Elegies* inhabit this non-dualistic space: not just children and animals, but also the young dead because they die before they have been corrupted by 'conformation' (*Gestaltung*), the patterns of perception and behaviour that separate us from our original home; lovers too have greater intensity than most of us:

> A child
> sometimes gets quietly lost there, to be always
> jogged back again. Or sometimes dies and *is* it.
> For, nearing death, one perceives death no longer,
> and stares ahead – perhaps with large brute gaze.
> Lovers – were not the other present, always

blocking the view! – draw near to it and wonder [. . .]
Behind the other, as though through oversight,
the thing's revealed . . . But no one gets beyond
the other, and so world returns once more.
Always facing Creation, we perceive there
only a mirroring of the free and open, dimmed by our breath.
. . . For this is Destiny: being opposite,
and nothing else, and always opposite.
(L. & S. 1963: 78–9)

    Als Kind
verliert sich eins im Stilln an dies und wird
gerüttelt. Oder jener stirbt und *ists*.
Denn nah am Tod sieht man den Tod nicht mehr
und starrt *hinaus*, vielleicht mit großem Tierblick.
Liebende, wäre nicht der andre, der
die Sicht verstellt, sind nah daran und staunen . . .
Wie aus Versehn ist ihnen aufgetan
hinter dem anderen . . . Aber über ihn
kommt keiner fort, und wieder wird ihm Welt.
der Schöpfung immer zugewendet, sehn
wir nur auf ihr die Spiegelung des frein,
vons uns verdunkelt. Oder daß ein Tier,
ein stummes, aufschaut, ruhig durch uns durch.
Dieses heißt Schicksal: gegenüber sein
und nichts als das und immer gegenüber.
(S. W. I 1955: 714f.)

    The Elegies lament man's failure to inhabit or even to appreciate the unity of world-inner-space, but they are not unalleviated sorrow. They alternate between lament and joy, each elegy presenting a different aspect of Rilke's diagnosis of man's alienation.
    The final impression left by the whole cycle is that we can find our true being, can find unity and fulfilment in the present moment and still have something to offer even the angels: we can contribute by praising life's potentiality through art. The Ninth Elegy puts it like this:

Here is the time for the tellable, *here* is its home,
speak and proclaim. More than ever
things we can live with are falling away, for that
which is oustingly taking their place is an imageless act.
Act under crusts, that will readily split as soon
as the doing within outgrows them and takes a new outline.

> Between the hammers lives on
> our heart, as between the teeth
> the tongue, which, in spite of all,
> still continues to praise.
>
> (L. & S. 1963: 85)

Hier ist des Säglichen Zeit, *hier* seine Heimat.
Sprich und bekenn. Mehr als je
fallen die Dinge dahin, die erlebbaren, denn,
was sie verdrängend ersetzt, ist ein Tun ohne Bild.
Tun unter Krusten, die willig zerspringen, sobald
innen das Handeln entwächst und sich anders begrenzt.
Zwischen den Hämmern besteht
unser Herz, wie die Zunge
zwischen den Zähnen, die doch,
dennoch, die preisende bleibt.

(*S. W. I* 1955: 718f.)

Despite our disinheritance from our true home in the 'Open', praise and 'transformation' remain our vital 'task', our '*Auftrag*'. 'Why', Rilke asks in the Ninth Elegy:

> oh, why
> *have* to be human, and shunning destiny,
> long for Destiny? . . .
>
> [. . .] because being here is much, and because all this
> that's here, so fleeting, seems to require us and strangely
> concerns us. Us the most fleeting of all. Just once,
> everything, only for once. Once and no more. And we, too,
> once. And never again. But this
> having been once on earth – can it ever be cancelled?
>
> [. . .] Are we, perhaps, *here* just for saying: House,
> Bridge, Fountain, Gate, Jug, Fruit tree, Window, –
> possibly: Pillar, Tower? . . . but for *saying*, remember,
> oh, for such saying as never the things themselves
> hoped so intensely to be.
>
> (L. & S. 1963: 83, 85)

>   warum dann
> Menschliches müssen – und, Schicksal vermeidend,
> sich sehnend nach Schicksal? . . .
>
> [. . .] weil Hiersein viel ist, und weil uns scheinbar
> alles das Hiesige braucht, dieses Schwindende, das

seltsam uns angeht. Uns die Schwindendsten. *Ein* Mal
jedes, nur *ein* Mal. Nie wieder. Aber dieses
ein Mal gewesen zu sein, wenn auch nur *ein* Mal:
irdisch gewesen zu sein, scheint nicht widerrufbar.

[. . .] Sind wir vielleicht hier, um zu sagen: Haus,
Brücke, Brunnen, Tor, Krug, Obstbaum, Fenster, –
höchsten: Säule, Turm . . . aber zu sagen, verstehs,
oh zu sagen, so wie selber die Dinge niemals
innig meinten zu sein.

(*S. W. I* 1955: 717f.)

Though Rilke is here focusing on the role of the poet, the sayer, he is also showing that all human beings, even if they are alienated from their true *being*, can find unity and raise themselves from suffering by fulfilling their task (*Auftrag*) which is: to praise simple, pure things before they were corrupted by mass production and the machine age, and to transform them into the '*invisible*' oneness of *Weltinnenraum*:

Praise this world to the Angel, not the untellable: you
can't impress him with the splendour you've felt; in the
   cosmos
where he more feelingly feels you're only a novice.
   So show him
some simple thing, fashioned by age after age,
till it lives in our hands and eyes as part of ourselves.
Tell him *things*. [. . .] These things that live on departure
understand when you praise them: fleeting, they look for
rescue through something in us, the most fleeting of all.
Want us to change them entirely, within our
   invisible hearts,
into – oh, endlessly – into ourselves! Whosoever we are.

Earth, is it not just this that you want: to arise
invisibly in us? Is not your dream
to be one day invisible? Earth! invisible!
What is your urgent command, if not transformation?

(L. & S. 1963: 87)

Preise dem Engel die Welt, nicht die unsägliche, ihm
kannst du nicht großtun mit herrlich Erfühltem im Weltall,
wo er fühlender fühlt, bist du ein Neuling. Drum zeig
ihm das Einfache, das, von Geschlecht zu Geschlechtern
   gestaltet,

> als ein Unsriges lebt, neben der Hand und im Blick.
> Sag ihm Dinge. [. . .] Und diese, von Hingang
> lebenden Dinge verstehn, daß du sie rühmst; vergänglich
> traun sie ein Rettendes uns, den Vertgänglichsten, zu.
> Wollen, wir sollen sie ganz im unsichtbaren Herzen
>   verwandeln
> in – o unendlich – in uns! Wer wir am Erde auch seien.
>
> Erde, ist es nicht dies, was du willst: *unsichtbar*
> in uns erstehn? – Ist es dein Traum nicht,
> einmal unsichtbar zu sein? – Erde! unsichtbar!
> Was, wenn Verwandlung nicht, ist dein drängender
>   Auftrag?
>
>   (*S. W. I* 1955: 719f.)

This difficult phrase 'transformation into the invisible' is clearer if, rather than merely understanding with our intellects the unity Rilke is striving for, we can experience it directly, face to face, pragmatically, just as the Buddha showed the way not through understanding but through living: through learning to live beyond universal suffering, freed from the pain of separation and reaction (*sankhāra*) by meditation. Rilke tried to explain transformation to von Hulewicz in a letter written on 13 November 1925:

> Transitoriness is everywhere plunging into profound Being. And therefore all the forms of the here and now are not merely to be used in a time-limited way, but, so far as we can, instated within those superior significances in which we share. Not, though, in the Christian sense (from which I more and more passionately withdraw), but, in purely terrestrial, deeply terrestrial, blissfully terrestrial consciousness, to instate what is here soon and touched within the wider, within the widest orbit – that is what is required. Not within a Beyond, whose shadow darkens the earth, but within a whole, within the Whole. (L. & S. 1963: 157)

It is we, Rilke tried to show, who, despite our limitations, can establish a link between the tangible and the intangible, between the everyday world of normal experience and the 'world-inner-space' of pure consciousness. Also in the famous Hulewicz letter he wrote:

> Nature, the things we associate with and use, are provisional and perishable; but so long as we are here, they are our possession and our friendship, sharers in our trouble and gladness, just as they have been the confidants of our ancestors. Therefore, not only must all that is here not be vilified or degraded, but, just because of that very provisionality they

share with us, all these appearances and things should be, in the most fervent sense, comprehended by us and transformed. Transformed? Yes, for our task is to stamp this provisional, perishing earth into ourselves so deeply, so painfully and passionately, that its being may rise again, 'invisibly', in us. We are the bees of the Invisible. *Nous butinons éperdument le miel du visible, pour l'accumuler dans la grande ruche d'or de l'Invisible.* (L. & S. 1963: 157)

Our task is thus to transform the divisions that we have created into the unity which is blissful and *invisible* because we are then within it: not outside looking in, or inside looking out; we are no longer separate, no longer 'opposite'. We are *present*, we are *here* both spacially and temporally; we can thus find ourselves by losing our selves.

The only maps we need are those which make themselves redundant, like the ladder which we kick away when we have climbed to our goal, or the raft we have used to cross to the other shore and seen that it was actually this shore all along. The *upāya* ('skilful means') thus become simply irrelevant, part of the former divided world, the multi-verse which we can transform into universe, world-inner-space, the Open, the Free, the Great All-Oneness of which we can become an integral part.

REFERENCES

Heidegger, M. (1955). *Language, Poetry, and Thought.* New York: Harper Row.

Leishman, J. B. and Spender, S. (1963). *Duino Elegies, the German Text with an English Translation, Introduction and Commentary.* London: Hogarth.

Rilke, R. M. (1955). *Sämtliche Werke* (2 vols). Frankfurt am Main: Rilke-Archiv.

DAVID MACLAGAN

Inner and Outer Space: Mapping the Psyche
For Simon Lewty

THE OBVIOUS AND ORDINARY FUNCTIONS OF MAPS

The two main functions of conventional maps of 'out-there' territories are those of orientation and navigation: the first being concerned with location (where are we?) and the second with relation (how do we get from A to B?). However, strictly pragmatic maps, whose functions are limited to these, form only a small proportion of the total body of maps, which ranges from casual sketch-maps to carefully contrived cosmological diagrams. Here we immediately come across the problem of what the difference might be between a map and a diagram: in theory, the former applies to an actual, tangible territory, while the latter refers to a more abstract or metaphysical one, to dynamics rather than to topography. However, a quick glance at the multiple and overlapping functions of maps soon makes such a neat distinction impossible to operate.

Maps serve a wide variety of functions, both practical and theoretical. They include the following:

Orientation. Navigation is not always as literal as it seems: a map is like a mnemonic device (this is much more evident in early European maps or in non-European maps): landmarks are also historical or mythical sites of memory.

Ordering. Maps establish location and relationship, not just of physical features, but of political boundaries, ethnic distribution, age, sex, religious belief, etc. In addition, plotting the maps of a territory is often a way of establishing a claim upon it, a kind of symbolic taking possession of it. Similarly, a map can justify a certain hierarchy (for example, in terms of upper and lower).

Instrument of knowledge. Knowledge itself is a form of appropriation (e.g. the naming of rivers, mountains, etc). Many maps (especially 17th century European ones) display and celebrate themselves as signs of the empire of scientific conquest (Mukerji 1984).

Analogical structure. A map is a analogue of certain features of a territory: its accuracy depends on a sufficient degree of correspondence with this territory (Korzybski 1948). But mapping also entails the superimposition of different sorts of information. This use of analogy can be metaphorical or conceptual, as well as functional: maps may correlate or cross-fertilise different domains, e.g. the signs of the Zodiac with the parts of the human body.

As an autonomous object. A map can be a virtuoso display of technical ingenuity or a work of art in its own right, independent of its accuracy (Mukerji 1984: 39–41).

As fantasy. Makers of fictional worlds (of which the 'paracosms' of childhood are prototypes [MacKieth 1983: 50–4]) create maps of their imaginary territories (e.g. Middle Earth or Nintendo maps). But maps are also made as fantasies *per se*: such maps may be completely invented, or they may be an undecidable mixture of fact and fantasy. These 'subjective' maps pose the problem of the nature of so-called 'inner space' and its relation to external reality.

PSYCHE AND ITS SPACES

By 'psyche' I mean the area of our experience that manifests itself in the world of feeling, memory and imagination (including fantasy and dreaming). A powerful and long-standing tradition in our culture locates this area in terms of an 'inner world'. Even if its main features are acknowledged to be shared by most people, the common assumption is that this is a largely invisible, subjective world, the intimate recesses of which are peculiar to each individual and essentially private.

There are two, mutually symmetrical, overlaps: one between private worlds and the public world of shared reality, and the other between the inner world and so-called objective reality (Lowenthal 1961). Just as from a phenomenological perspective no shared public world is entirely free from subjective elements, so our 'private worlds' contain many elements that have been imported from external perception. Many models of the mind (for example, Freud's) assume that *all* the ingredients of fantasy or imagination are regurgitations of elements from previous perceptual experience. But some other models (for example, Jung's)

Inner and Outer Space: Mapping the Psyche 153

propose that psychic life is only *apparently* confined within the individual and that some of its constituents are reflections of a transpersonal psychic force-field (archetypes, for example).

As objective reality has come to be identified with the external world, and the non-visible world of psychic experience has been increasingly categorised as unreal or subjective, so the question of how this inner world is to be translated into the borrowed idiom of perceptual reality (e.g. in paintings or films about dreams or visions) has become more difficult (Maclagan 1989). Thus there is an influential cultural habit that leads us to construct inner or psychic space in terms of external space, modelled in terms of Euclidean co-ordinates. When we inhabit this space imaginatively, the assumption often made is that we live and move in it by analogy with external perception: for example, that what we 'see with the mind's eye' is not essentially different from what we see with our outward eye. It is questionable how far this picture corresponds with the actual phenomenological description of inner space. But the situation is further complicated by the fact that cultural images (texts, paintings and, above all, moving pictures) don't just document this space but are also fed back, incorporated into the idiom of our fantasy, so that they inform or construct it.

What alternative psychic spaces can we conceive of? There are forms of psychic experience on record that don't fit these conventional pictures: for example, mediumistic drawings or psychedelic imagery, and perhaps some works by 'psychotic artists'. These suggest a very different experience of spatial or temporal location (or dislocation), in which there is something like a collapse or implosion (Fischer 1971: 897–903). But however abstract these images may appear, they are still *pictures*: their relationship to a diagram of the psyche is like that of a landscape painting or photograph to a map of the territory of which it is a partial view, from a more local perspective. Or they might be more like sketch-maps or plans of a personal domain than large-scale psychic geographies. Like other maps, maps of inner space can take an informal and subjective, or a more formal or abstract form, depending on what level of experience they are dealing with.

Objective maps of psyche

It may be that the more inner or psychic space is conceived of by analogy with external perception (as an inner vision, a private theatre or a movie in our heads), the more disembodied and abstract are the models or maps that we make of it. The more objective of these maps have an extremely remote, generalised and diagrammatic character: for example, they portray

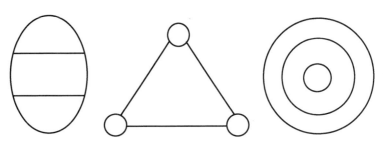

Figure 1   Different diagrammatic ways of mapping three elements give different pictures of their interrelationship.

psychodynamics in terms of a topographical structure whose co-ordinates (upper and lower levels, the inclusion of one domain within another, the reciprocity or assymetry of energetic movements, etc.) establish both a causal mechanics and hierarchy of values. An obvious example is the inferior position of the unconscious in Freud's model of the psyche and the distortion of communication between it and consciousness that results from the horizontal barrier of repression.

Such psychological maps of psyche are speculative or didactic: they also serve the purpose of claiming a territory by naming it: they are not used for 'navigation' in inner space, nor do they refer to particular or identifiable features. Indeed, the whole question of the extent to which they can be verified is problematic, since the boundaries and topography of the features they indicate are largely indeterminate, and their status very much a matter of belief or assertion.

In such an abstract and prospective domain, the question inevitably arises as to what extent the match between a map and a supposed territory is a matter of invention rather than fidelity. Formal and symmetrical maps will tend to generate their own assumptions. For instance, in modelling the relation between three functions, we might choose any of the following forms: a tripartite egg (divided vertically or horizontally into three); a series of three concentric rings; or a triangle – to name only the most obvious options. Each of these models gives a different picture, a different dynamic relationship, a different hierarchy. Jung's preoccupation with four-fold structures, which he applied to psychological functions, the structure of the self and to alchemical imagery (Jung 1959: 226–65), is a striking example of this reciprocal relationship between a map and a territory. Clearly, he believed that quaternity was an archetypal ordering principle (Von Franz 1974) and the diagrams he composed are intermediary between the metaphysical realm of the 'psychoid aspect' of archetypes and its eventual manifestation in the phenomenal world: to what extent these schemes of correspondence can be described as 'maps' is another question.

Subjective maps of psyche

Objective maps tend to use the regular geometric structures – such as grids or axes of symmetry – that are characteristic of generalised systems, and that can therefore be presumed to refer to collective rather than personal forms of experience. Such 'scientific' visual conventions are one of the pictorial idioms that best qualify to be described as a 'language', since they are both written and read according to rules established in advance. However, there are a few examples of maps that, while they use this objective language, do so for quite personal and subjective purposes: indeed the reading of such maps is more a matter of symbolism or metaphor than of literal location. There is an ironic or paradoxical twist in using such an impersonal means to hint at highly personal or secret meanings (Roux and Hubert 1983: 2223–34).

More obviously, subjective maps refer directly both to specific features of personal experience – feelings, dream associations – and also to the private dimension of memory. The effect of such maps, whether intended or not, is to re-introduce many of those features of experience, labelled as subjective and assumed to be interior, that are excluded from conventional 'scientific' maps. Some of the key characteristics of these maps include:

- the treatment of feelings, states of mind or psychic events (such as dreams) on the same footing as physical features or concrete facts;
- the use of a flexible framework or scale, rather than a uniform one;
- the adoption of a frankly idiosyncratic or personal perspective, i.e. subjective experience is the lens through which the convergence of phenomena is registered.
- the inclusion of time (in the form of memory) and motion (in the form of narrative) in the map.

These kinds of subjective map could be seen as establishing the basis for a personal mythology or a private cosmology – i.e. something that transcends the boundary between interior and exterior experience. What does this mean in practice? It may mean treating certain of the accidental circumstances of one's life as landmarks, giving them a peculiar prominence, so that a cluster of associations can be attached to them: they thus acquire metonymic

Figure 2 Adolf Wölfli, the Swiss 'psychotic artist', was interned in the Waldau asylum for the last thirty-five years of his life (1895–1930). From within this confinement he created a vast body of work that reproduces the world of external scientific and technical knowledge (botany, geography, mechanics, architecture) in a systematic but highly idiosyncratic idiom. The imaginary journeys he described were illustrated with an amalgam of pictographic, alphabetical, ornamental and musical signs, and included a number of maps inspired by the atlas, which was one of the few books available to him in the asylum (*London-Nord*, 1911).

Inner and Outer Space: Mapping the Psyche   157

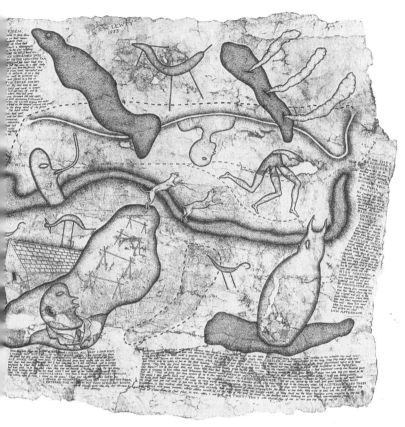

Figure 3   Much of the artist Simon Lewty's work revolves around the various ways in which the internal 'landscapes' of memory or dream overlap with external landscapes. His paintings deploy a wide range of inscription, some of them literal and some corresponding to a wider sense of 'writing' (écriture): they range from ancient to modern, from graffiti to Ordnance Survey conventions. Both the 'narrative' of his texts and the 'topography' of the fields that incorporate them constitute a no-man's-land between history and fiction, public and private worlds (*The Beginning of Ridicule*, 1983: The Bank of England).

or metaphorical functions (perhaps one could call this the 'madeleine syndrome'). Or it may mean squatting in the ruins of previous maps, creating a bricolage out of their debris, or giving it a new twist. This sometimes amounts to another form of subjective mapping: that of setting up correspondences between one system and another, or of turning all the properties of one concrete object into symbols for another, intangible entity (e.g. a paraffin stove as model of the psyche). Here the heuristic or prospective aspect of analogy comes into play: trying out a superimposition, the full import of which has yet to be disclosed, to 'see what happens'. Both the mythologising and analogising forms of mapping have the advantage over more abstract diagrammatic forms in that they relate the specifics of psychic experience to more general psychological landscapes, in ways that resemble pre-scientific maps. That is why I believe that the most interesting – and maybe the most informative – maps of psyche are likely to be provided by artists rather than by psychologists.

REFERENCES

Fischer, R. (1971). A Cartography of the Ecstatic and Meditational States. *Science* vol. 174, no. 4012: 897–903.

Jung, C. G. (1959). *Alion*. London: Routledge and Kegan Paul.

Korzybski, A. (1948). *Selections from 'Science and Sanity'*. Connecticut: International Non-Aristotelian Library Publishing Co.

Lowenthal, D. (1961). Geography, Experience and Imagination. *Annals of the Association of American Cartographers*, vol. 51, no. 3.

MacKeith, S. A. (1983). Paracosms and the Development of Fantasy in Childhood. *Ambit* 100: 50–4.

Maclagan, D. (1989). Fantasy and the Figurative. In Gilroy, A. and Dalley, T. (eds) *Pictures at an Exhibition*. London: Routledge.

Mukerji, C. (1984). Visual Language. *Science and the Exercise of Power: the Case of Cartography in Early Modern Europe*. In *Studies in Visual Communication*, vol. 10, no. 3.

Roux, G. and Hubert, M. (1983). Alain le Cartographe. *Psychologie Medicale*, vol. 15, 13: 2223–34.

Von Franz, M-L. (1974). *Number and Time*. London: Rider.

BRIAN BOCKING

'If you meet the Buddha on the map ...':
The Notion of Mapping Spiritual Paths

This chapter is about the notion of mapping spiritual paths, my thesis being that spiritual maps (maps of spiritual paths) are different in kind from ordinary maps. Another way of putting this is to say that a spiritual map is a map which differs from ordinary maps in the ways indicated below. Reading spiritual maps as if they were ordinary maps I would characterise as MAP or 'Mysticism for Academic Purposes'. MAP-ping is an interesting intellectual activity which is sometimes confused with mysticism itself.

THE GNAT AND THE WIND

In book three of the Mathnavī of Jalālu'ddin Rūmī, there is the story of a gnat appealing to Solomon for justice against the Wind whose strength oppresses him. Solomon agrees to hear the case provided that both parties are present, for 'until both litigants come into the presence, the truth does not come to light before the judge'. As soon as the Wind arrives at court, the gnat is blown away. Commenting on this parable, Rūmī says, 'Even such is the seeker of the Court of God: when God comes, the seeker is naughted' (Nicholson 1930: 258–60).

The gnat and the Wind cannot be present together, for where one is, the other is not. Even Solomon cannot do them both justice. Their coexistence cannot be mapped.

The Zen saying from which the title of this paper is derived runs, 'If you meet the Buddha on the road – kill him!' This means, do not be obstructed in spiritual progress by your limited conceptions of what Buddhism is. The spiritual teachings of many traditions include reminders that the true nature of things cannot be grasped by the intellect (that is, mapped). Buddhist teachings have an explicit concern with the relationship between the ordinary effort to map reality (what we call scientific accounts of the nature of the world) and the rather different activity of producing and following spiritual maps (spiritual teachings). The Buddha applies to his own teachings the simile of a raft: once the river is crossed, the raft is abandoned. The Buddha's teachings are to be used for a particular purpose and then completely discarded, unlike ordinary theories about the nature of the world, which aim to be enduring. The raft analogy reminds the practitioner that Buddhist concepts (such as non-self) are useful, like a raft, but just as reaching dry land renders the raft redundant, so the (actual) attainment of non-self implies the extinction of conceptions of both self and non-self. In Rūmī's Sufi example, merging with God extinguishes thoughts of both self and God (the gnat is blown away entirely).

ORDINARY MAPS

An ordinary map is an accurate record or representation of a terrain. For example, Marco Polo goes to China, and produces a map which others can follow if they want to go there. In practice, however, an ordinary map is a very limited device because terrain changes, and maps are typically produced in the past to be used in the future.

Also, an ordinary map is static – it is a 'snapshot' of a terrain at one point in time. A map also necessarily shows some features but not others: the road, for example, but not where the bandits are today (or where the bandits are, but not where your vehicle is going to break down). To be a sufficient guide, an ordinary map needs updating by methods which include:

1 collecting additional information before departure (creating a new map);
2 acquiring supplementary information along the way;
3 taking as a guide one who has already made the journey;
4 relying on instinct (homing pigeons do not need maps).

Sophisticated 'ordinary' maps, such as the television weather map, are constantly updated in a variety of ways. Nevertheless such maps remain 'snapshots'. They are always out of date, and their predictive value is not 100 per cent.

SPIRITUAL MAPS

Spiritual maps, unlike ordinary maps, are evaluative and hierarchical. An ordinary map may show that all roads lead to Rome but it does not thereby imply that one should go to Rome. A spiritual map shows where we are now, and where we should be; such a map contains an imperative to transcend the present location; the 'places' on the map embody an implicit hierarchy. Samsara and nirvana on the Buddhist spiritual map are not like North and South on an ordinary map; there is in the Buddhist case an implicit instruction or exhortation to move from one to the other.

Because spiritual maps are hierarchical, they incorporate their own redundancy. A spiritual map which charts the transformation of lust into divine love, or egotism into union with the Lord, or samsara into nirvana, apparently showing them both on the same 'map', is actually a projection of the spiritual path as seen from the start of the journey. The teaching addressed to the spiritual seeker at the start of the journey is not a description of the way things really are.

When we travel from London to Lhasa, London does not cease to exist because we have reached Lhasa. Both remain on the map. A spiritual journey is something different, for 'when the light dawned the darkness vanished' (Nicholson 1930: 259). The attainment of nirvana in Buddhism implies the complete cessation of samsara – samsara is no longer on the map. Union with God means that separation from God is no longer a feature of the terrain. A spiritual map, in other words, represents a process of refinement and transformation, rather than a straightforward journey from A to B. At each stage, were it to be mapped, the map of the remaining journey would become different.

Buddhist thought makes this self-negating characteristic of spiritual maps explicit in acknowledging that the Buddha provided a useful account of the spiritual path in the form of conventional Buddhist teachings about non-self, impermanence, the five skandhas, Mount Sumeru, etc. out of his compassion for living beings, while at the same time asserting that this map should not be grasped as a description of how things really are. It is a teaching by the Buddha which deals skilfully with the ignorant projections of unenlightened minds. To illustrate this point, a parable is told in the Lotus Sutra of a magic city, conjured up by a guide when travellers in a desert begin to lose heart on their journey. Encouraged by the appearance of the illusory city, the travellers hasten forward, thereby making progress on the real journey. The city is there on the map, but it is not there in reality; the travellers are making a journey, but it is not the journey they think they are making. The real journey is unmapped, because the travellers would never embark on the journey if they understood what it really entailed. In the same way, the gnat would never have asked Solomon for justice, if it had understood the real nature of the Wind.

The self-negating character of spiritual maps implies that a spiritual map is good for only a very tiny part of the spiritual journey – perhaps no more than the initial satisfaction of intellectual curiosity and the settling of doubts. As soon as any spiritual progress is made, the map changes, as the Lotus Sutra parable suggests. The temporary character of spiritual maps underlines the need for a personal guide on the spiritual path, a need which is emphasised in all spiritual traditions but which tends to be overlooked in purely intellectual MAP-ping, for all sorts of post-Enlightenment reasons.

FROM 'A' TO 'B'

In summary, a spiritual map may appear to be a representation of a journey from A to B, but the map of the journey, being a projection tailored by the teacher to the limited standpoint of the disciple at A, does not represent the real spiritual journey, though it has to be thoroughly convincing from the standpoint of A. Even the description of 'A' (e.g. that we are unenlightened) is an adaptation to the standpoint of one at A. With each step on the spiritual path perception changes and the map that initiated this particular step becomes redundant. This means that the spiritual journey is largely uncharted after the first step, which is why a spiritual guide is essential.

The goal of the spiritual path may appear on the map as 'B' (e.g. enlightenment) but the true reality of B cannot be represented on the same map as A. It is conventionally represented as 'B', to accommodate the standpoint of A, but 'B' on the map is not what B really is.

In Rūmī's parable, A is the gnat (the self), and B is the Wind (God). In theory, with the wisdom of Solomon, they can be brought together in harmony (occupying the same map), but once the two are actually brought together (once the self begins to approach God) it becomes clear that gnat and Wind, ego and God, are mutually exclusive. 'In this place of presence', says Rūmī, 'all minds are lost beyond control; when the pen reaches this point, it breaks'.

REFERENCES

Nicholson, R. A. (1982). The Mathnawī of Jalālu'ddin Rūmi, vol. 4. Cambridge: Gibb Memorial Trust.

Notes on Contributors

BRIAN BOCKING is head of the department of Study of Religions at Bath College of Higher Education. He was formerly lecturer in Japanese Religions at the University of Stirling, Scotland, and visiting lecturer at the University of Tsukuba, Japan. His research interests are Sino-Japanese Buddhism, Japanese religions and comparative religions. His publications include *Nagarjuna in China: a Translation and Study of the Middle Treatise* (1993), Mellen Press; 'RAP, RFL and ROL: Language and Religion in Higher Education', in Masefield and Wiebe (eds) *Aspects of Religion: Essays in Honour of Ninian Smart* (1993); and 'The Origins of Japanese Philosophy', in Mahalingam and Carr (eds) *Encyclopaedia of Asian Philosophy*.

HILDA R. ELLIS DAVIDSON was formerly President of the Folklore Society and Fellow and Vice-President of Lucy Cavendish College, Cambridge. She is author of *Gods and Myths of Northern Europe* and a number of other books and articles on Norse mythology and early religion in western Europe.

GAVIN D. FLOOD is lecturer in Religious Studies in the Theology and Religious Studies Department, University of Wales, Lampeter. His research interests are in the areas of Indian philosophy and religion, ritual and methodology. His publications include *Body and Cosmology in Kashmir Śaivism* (1993), Mellen Press, and papers on Kashmir Śaivism in *Numen* and *Religion*.

JOHN M. FRITZ teaches in the Department of Anthropology, University of New Mexico, Albuquerque. His publications include 'Was Vijayanagara a Cosmic City?', in A. L. Dallapiccola (ed.) *Vijayanagara – City and Empire: New Currents of Research* (1985); 'The Plan of Vijayanagara', in A. L. Dallapiccola (ed.) *Shastric*

Traditions in Indian Art (1989); and, with G. Michell and M. S. Nagaraja Rao, *Where Kings and Gods Meet, The Royal Centre at Vijayanagara* (1984), University of Arizona Press.

GRAHAM HARVEY lives in Tynemouth and is engaged in post-doctoral research into contemporary Paganism and nature spiritualities and responses to them.

EMILY LYLE is a research fellow at the School of Scottish Studies, University of Edinburgh. She has held fellowships at the Radcliffe (now Bunting) Institute, Harvard University and the Humanities Research Centre, The Australian National University, and has lectured on folklore and folklife at Stirling and UCLA. Her publications in the fields of folksong, folklore, ethnology and cosmology include *Archaic Cosmos: Polarity, Space and Time* (1990), Polygon.

DAVID MACLAGAN is a lecturer at the Centre for Psychotherapeutic Studies, University of Sheffield. He is a writer, artist and art therapist. He is author of *Creation Myths* (1977, Thames and Hudson), and of numerous articles on Outsider Art and on the overlap between psychological and aesthetic qualities. His interest in 'psychic maps' goes back to the late 1970s and to his work as a performance artist and in participatory seminars on Maps and Mapping, given at St Martin's School of Art and Maidstone College of Art.

J. McKIM MALVILLE is a Professor of Astronomy at the University of Colorado at Boulder. His research interests have included the *aurora australis* (he spent winter in the Antarctic during the International Geophysical Year), the sun – such as sunspots, flares, and its corona – and, most recently, the archaeoastronomy of India and Anasazi ruins of the American South-west. His books include *A Feather for Daedalus: Explorations in Science and Mythology* (1975), *The Fermenting Universe: Myths of Eternal Change* (1981) and *Prehistoric Astronomy in the Southwest* (1989).

CHRIS MORRAY-JONES is currently a visiting scholar at Stanford University. He graduated from Manchester University with a BA in Comparative Religion and has a PhD from Cambridge University. From 1989 to 1992 he was the Gordon Milburn Junior Research Fellow in the Study of Mysticism and Religious Experience at Lady Margaret Hall, Oxford. He has published articles in the *Journal for the Study of Judaism* and *The Journal of Jewish Studies*, and he is writing a book on the influence of Jewish esoteric mysticism on the authors of the New Testament.

Notes on Contributors

MARK NUTTALL is lecturer in Social Anthropology at Brunel University. He has an MA from Aberdeen University and a PhD from Cambridge. Between 1990 and 1992 he was Research Fellow in Social Anthropology at the University of Edinburgh. He has done fieldwork in Scotland and, for two years, in Greenland.

WENDY PULLAN was a lecturer in design and Architectural History and Theory in the Department of Architecture and Environmental Design at Bezalel Academy, Jerusalem, from 1979 to 1989. She has written *A Guide to the Architecture of Jerusalem*, as well as a number of articles on the city. At present she is a Research Fellow at Cambridge University and is completing a PhD there on the nature of sacred space in early Christian Jerusalem.

URSZULA SZULAKOWSKA is an art critic and senior lecturer in Art History and Theory at Bretton Hall College, University of Leeds. She has published work on Renaissance and twentieth century alchemical illustration in *Ambix* (1986, 1988), *Cauda Pavonis* (1988, 1992), *Chryspeia* and *Acta Historia Artium* (1993). She is writing a book on alchemy and sexuality in art.

CYRIL WILLIAMS is Professor Emeritus, University of Wales. He was formerly Professor of Religious Studies at St David's University College, Lampeter; Professor of Religion at Carlton University, Ottowa; President of the British Association for the History of Religions; Dean of Divinity, University of Wales; and Fellow of the University of Wales, Cardiff. His publications include *Tongues of the Spirit* (1981) University of Wales; *Y Fendigaid Gan* (1991) University of Wales, the first Welsh translation of the *Bhagavad Gita*; and *Basic Themes in the Comparative Study of Religion* (1992), Mellen Press.

ROY WOODS teaches German language, literature and film (including a course on subtitling) at St David's University College, Lampeter. He has written an MA thesis on the problems of translating Rilke's poetry into English and is interested in both theoretical and practical aspects of oriental philosophy and religion. He is a practising *vipassana* meditator. Relevant publications include 'Some Aspects of Seeing in Rilke's Poetry', *Trivium* 8 (1973): 95–108; 'Echoes of Rilke', *Trivium* 27 (1992): 85–196; 'The Fasting Mind, a Syncretic Approach to Meditation', *Studia Mystica* 14 (1991): 3–21; and 'Death and Ontology in Rilke', *Trivium* 27 (1992): 109–42.

Cosmos is the yearbook of the Traditional Cosmology Society which is concerned with exploring myth, religion and cosmology across cultural and disciplinary boundaries, and with increasing our understanding of world views in the past and present. For details of the society's activities and current subscription rates, please write to:

> Mrs Mary Brockington
> Secretary, Traditional Cosmology Society,
> 3 Eskvale Court
> Penicuik
> Midlothian
> Scotland EH26 8HT

Previous issues of *Cosmos* are available from:

> Edinburgh University Press Ltd,
> 22 George Square,
> Edinburgh,
> Scotland EH8 9LD

at the prices stated below.

1	1985	Duality	£12.50
2	1986	Kingship	£12.50
3	1987	Analogy	£12.50
4	1988	Amerindian Cosmology	£25.00
5	1989	Polytheistic Systems	£25.00
6	1990	Contests	£25.00
7	1991	Women and Sovereignty	£25.00
8	1992	Sacred Architecture in the Traditions of India, China, Judaism and Islam	£25.00

GENERAL THEOLOGICAL SEMINARY
NEW YORK

DATE DUE

MAR 1 5 2005			

HIGHSMITH #45230 Printed in USA